# DEVELOPMENTS IN ORIENTED POLYMERS—2

# CONTENTS OF VOLUME 1

## Edited by I. M. Ward

# DEVELOPMENTS IN ORIENTED POLYMERS—2

*Edited by*

## I. M. WARD

*Department of Physics,*
*University of Leeds, UK*

ELSEVIER APPLIED SCIENCE
LONDON and NEW YORK

ELSEVIER APPLIED SCIENCE PUBLISHERS LTD
Crown House, Linton Road, Barking, Essex IG11 8JU, England

*Sole Distributor in the USA and Canada*
ELSEVIER SCIENCE PUBLISHING CO., INC.
52 Vanderbilt Avenue, New York, NY 10017, USA

WITH 28 TABLES AND 144 ILLUSTRATIONS

© ELSEVIER APPLIED SCIENCE PUBLISHERS LTD 1987
Softcover reprint of the hardcover 1st edition 1987

**British Library Cataloguing in Publication Data**

Developments in oriented polymers.—
(Developments series)—2-
1. Polymers and polymerization—
Periodicals
I. Series
547.7        QD380

**ISBN-13: 978-94-010-8033-0        e-ISBN-13: 978-94-009-3427-6
DOI: 10.1007/ 978-94-009-3427-6**

# PREFACE

The last four years since the publication of the first of this series have seen further striking developments in both the science and technology of oriented polymers. In particular, polymers possessing very high degrees of molecular orientation are now quite commonplace, and this is reflected by the inclusion of five chapters dealing with ultra-high modulus polyethylene fibres, oriented liquid crystalline polymers (both lyotropics and thermotropics) and polydiacetylene single crystal fibres. At the same time there is continuing interest in the structure and properties of less highly oriented polymers and in the mechanisms of deformation in polymers. It is therefore good to have these themes represented also.

I should like to thank the contributors for adhering to a rather tight time schedule, and the publishers for their cooperation, so that this book can provide up-to-date reviews of the state of the art in a rapidly moving area of polymer science.

I. M. WARD

# PREFACE

# CONTENTS

# LIST OF CONTRIBUTORS

A. BISWAS

    *Department of Macromolecular Science, Case Western Reserve University, Cleveland, Ohio, USA.*

J. BLACKWELL

    *Department of Macromolecular Science, Case Western Reserve University, Cleveland, Ohio, USA.*

A. CIFERRI

    *Istituto di Chimica Industriale, University of Genoa, Genoa, Italy.*

M. G. DOBB

    *Textile Physics Laboratory, Department of Textile Industries, University of Leeds, Leeds LS2 9JT, UK.*

D. J. JOHNSON

    *Textile Physics Laboratory, Department of Textile Industries, University of Leeds, Leeds LS2 9JT, UK.*

R. KIRSCHBAUM

    *DSM, Central Research, Geleen, The Netherlands.*

P. J. LEMSTRA

    *Department of Polymer Technology, Eindhoven University of Technology, Eindhoven, The Netherlands.*

LUCIEN MONNERIE

  *Laboratoire de Physicochimie Structurale et Macromoléculaire, Ecole Supérieure de Physique et de Chimie Industrielles de la Ville de Paris, 10 rue Vauquelin, 75231 Paris Cedex 05, France.*

T. OHTA

  *Toyobo, Katata Research Institute, Ohtsu, Shiga, Japan.*

A. J. OWEN

  *Institut für Physik III, Angewandte Physik, Universität Regensburg, Universitssstrasse 31 Postfach, 8400 Regensburg, Federal Republic of Germany.*

H. YASUDA

  *Toyobo, Katata Research Institute, Ohtsu, Shiga, Japan.*

ROBERT J. YOUNG

  *Department of Polymer Science and Technology, UMIST, PO Box 88, Manchester M60 1QD, UK.*

*Chapter 1*

# POLYMER SINGLE CRYSTAL FIBRES

ROBERT J. YOUNG

*Department of Polymer Science and Technology,
UMIST, Manchester, UK*

## 1. INTRODUCTION

Historically most of the effort in obtaining highly oriented polymer samples has been concentrated upon trying to uncoil and align long polymer molecules in the solid state, melt or solution. A variety of techniques have been employed with conventional polymers, e.g. polyethylene, such as ultra-drawing,[1-3] high degrees of extrusion[1-3] and gel or solution spinning.[6,7] New rigid-rod polymers have also been developed from which highly oriented fibres can be prepared by the spinning of liquid crystalline solutions.[8-10] Although significant improvements and developments have taken place over recent years and samples with high values of modulus and strength have been prepared, perfect molecular alignment as in single crystals is never achieved using such approaches since defects such as chain-ends, chain-folds, loop and entanglements are invariably trapped in the structures.

This present chapter is concerned with an entirely different approach to the preparation of oriented high modulus polymers, namely solid-state polymerization. The technique has the ability to produce polymer single crystal fibres in which the molecules are perfectly aligned and the fibres have their theoretical values of stiffness and strength. The technique is essentially very simple. Single crystals of monomer molecules are prepared using conventional methods such as vapour phase deposition or precipitation from dilute solution. Alignment of the monomer molecules is present in the monomer crystals. The transformation from monomer to polymer takes place through a solid-state polymerization reaction by a rearrangement of bonding within the crystals but without any appreciable molecular movement. The alignment of the monomer molecules is therefore transferred to a uniaxial orientation of the polymer molecules. Any of the normal problems encountered with aligning polymer molecules in the melt or solution, such as entanglements, are therefore avoided. In this way polymer molecules can be incorporated into fibres without passing through the melt or solution stage which invariably produces polycrystalline samples and imperfections in the fibres.

Although the solid-state polymerization technique is simple and attractive, it suffers from several drawbacks. As yet, the fibres which are produced relatively slowly are only made as short strands with maximum lengths of no more than 5–10 cm. Also the only really successful single crystal fibre system is based on the solid-state polymerization of certain substituted diacetylenes,[11–13] and it cannot be applied widely to all polymer systems. Nevertheless, these substituted diacetylene polymer fibres are found to have promising and useful physical properties characteristic of one-dimensional solids. Although this chapter is concerned principally with the mechanical properties of polydiacetylene single crystals, it should be remembered that they also have unusual optical[14,15] and electronic properties[16,17] and have recently been described as 'prototype one-dimensional semi-conductors'.[18] The development of polydiacetylene single crystals has led to significant improvements in our understanding of many structure/property relationships in one-dimensional solids.

## 2.  SOLID-STATE POLYMERIZATION

### 2.1. Topochemical and Topotactic Polymerizations
Solid-state polymerization reactions have been discussed in detail by Wegner,[11] and his terminology will be used in this chapter. These reactions which are relevant to the preparation of polymer single crystal fibres are termed either 'topochemical' or 'topotactic' depending upon the detailed nature of the reaction. These two terms are used to describe reactions which can take place in organic crystals such as single crystals of fibre-forming monomers. In such crystals the molecules are generally separated sufficiently far that they are unable to react. However, if there is sufficient mobility that the molecules are able to come within about 0·3 nm of each other by diffusion or rotation, a solid-state reaction may occur.

Examples of such systems in which monomer single crystals can undergo solid-state polymerization have included (i) oxacyclobutanes such as trioxane,[19,20] (ii) diolefins undergoing 'four-centre type' photopolymerization,[21] and (iii) monomers with conjugated triple bonds such as substituted diacetylenes.[12,13]

It is found that only the second two examples are true topochemical reactions. In the case of trioxane, polymer crystals form inside the monomer crystals but, although there are crystallographic orientation relationships between the monomer and polymer crystals, the resulting samples of polyoxymethylene are polycrystalline with fibrillar polymer crystals[19] and consequently do not have very impressive mechanical properties.[20] The most important reactions for the formation of good polymer single crystals are topochemical reactions whereby, as Wegner[11] pointed out, 'there is a direct transition from the monomer molecules to polymer chains without destruction of the crystal lattice and without the formation of non-crystalline intermediates'. Hence the behaviour is rather like a martensitic transformation;[22] it is illustrated in Fig. 1 for the solid-state polymerization of a substituted diacetylene. The centres of gravity of the molecules do not move significantly from their lattice site and the reaction takes place within the parent crystal. The three-dimensional order of the monomer lattice is transferred directly to the polymer lattice with

TABLE 1

CHEMICAL FORMULAE AND ABBREVIATIONS FOR THE DIACETYLENE DERIVATIVES DESCRIBED IN THE TEXT WITH GENERAL FORMULA $R-C \equiv C-C \equiv C-R$

| R | Chemical name | Abbreviation |
|---|---|---|
| $-CH_2OCONHC_2H_5$ | 2,4-Hexadiyne-1,6-diol bis(ethyl urethane) | EUHD |
| $-CH_2OCONHC_6H_5$ | 2,4-Hexadiyne-1,6-diol bis(phenyl urethane) | PUHD |
| $-CH_2N$ | 2,4-Hexadiyne-1,6-di(N-carbazol) | DCHD |
| $-CH_2OSO_2C_6H_4CH_3$ | 2,4-Hexadiyne-1,6-diol bis(p-toluene sulphonate) | TSHD |

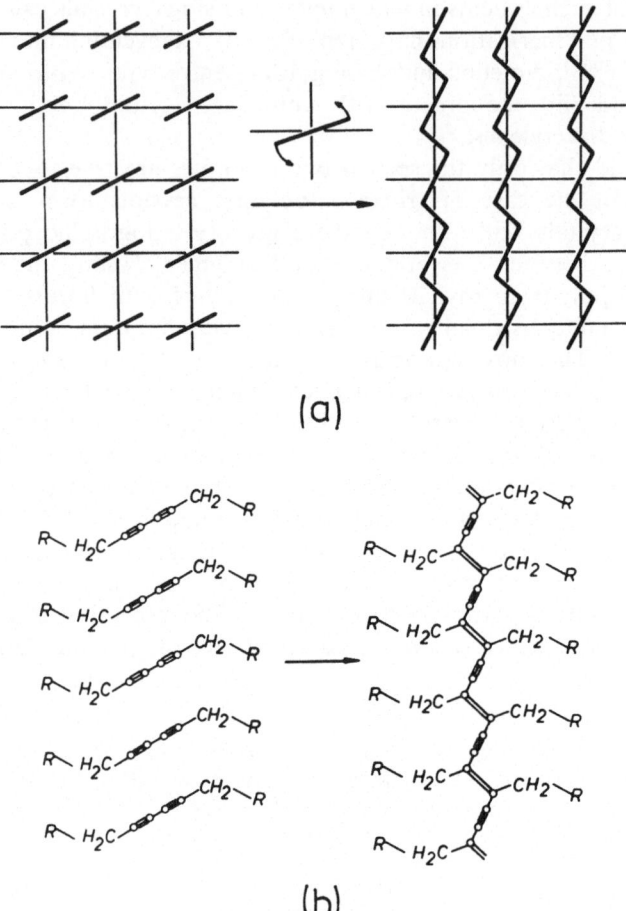

**FIG. 1.** (a) Schematic representation of a topochemical polymerization reaction. (b) Polymerization of a substituted diacetylene. (After Reference 11; reproduced with permission.)

the result that the polymer formed must have a well ordered extended-chain morphology. This behaviour was first reported by Wegner[12] who produced virtually defect-free extended-chain single crystals of substituted polydiacetylenes from single crystal monomers. In the intervening years a large number of different substituted diacetylenes have been produced; a number of examples are shown in Table 1 along with the abbreviations that will be used to describe them in this chapter.

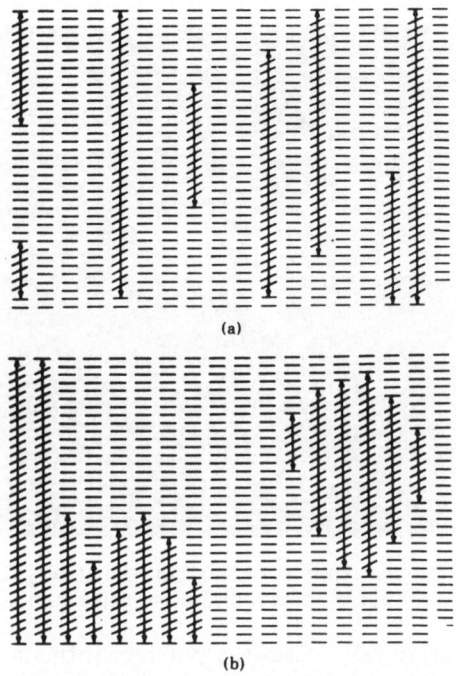

(a)

(b)

FIG. 2. Schematic representation of the formation of polymer molecules during topochemical polymerization: (a) homogeneous; (b) heterogeneous. (After Reference 11; reproduced with permission.)

## 2.2. Reaction Mechanisms and Kinetics

Topochemical polymerization reactions can proceed in either a homogeneous or a heterogeneous manner as shown in Fig. 2. In the homogeneous case (Fig. 2(a)) the polymer molecules form randomly in the monomer matrix. The other possibility is that polymerization is heterogeneous, starting at specific sites in the monomer crystal such as at defects (Fig. 2(b)). If further chain growth takes place at the pre-existing polymer nuclei there can be a tendency for the polymers formed to be polycrystalline as is found with four-centre type polymerization.[21] In the case of polydiacetylenes both homogeneous and heterogeneous polymerizations have been reported. For example, in the case of the toluene sulphonate derivative (TSHD) polymerized using synchrotron radiation, the reaction has been observed to take place homogeneously throughout a crystal,[23] whereas when TSHD monomer is polymerized thermally the polymerization reaction is found to start preferentially at

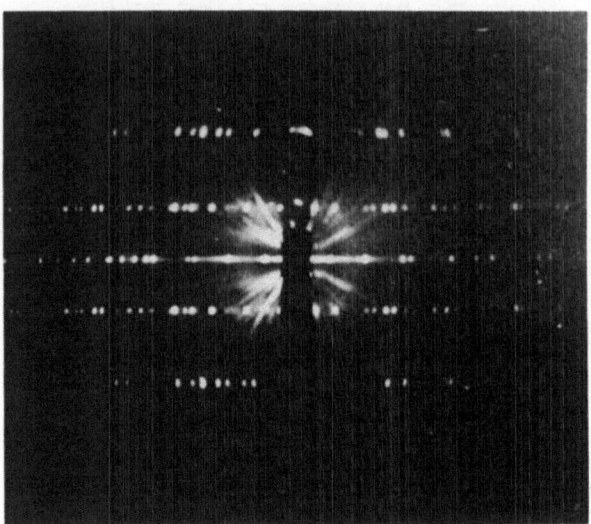

Fig. 3.   X-ray rotation photograph of a single crystal of polyTSHD rotated about
the chain axis; the sharp spots indicate the perfection of the crystal.

defects.[24] However, in both cases the polymer forms as a solid solution
with monomer, and the single crystal morphology is directly transferred
from monomer to polymer leading to polyTSHD crystals having a high
degree of perfection as can be seen from the rotation photograph in Fig. 3.

The behaviour of other diacetylene derivatives can be more complex
such as in the case of fibre-forming diacetylenes such as the carbazolyl
(DCHD)[25,26] and ethyl urethane (EUHD)[27–29] derivatives. In both cases
thermal polymerization produces polycrystalline samples which are not as
crystallographically    perfect    as    the    single    crystals    produced    when
polymerized using $\gamma$-rays at room temperature (cf. Fig. 3). It has been
suggested[25] that in the case of DCHD this may be due to the tendency for
phase separation to be favoured kinetically at the higher temperatures
used for the thermal polymerization.

Several mechanisms have been put forward to explain the formation of
polymer chains.[11] The most generally accepted one uses evidence from
ESR spectroscopy[30] where it is found that the reactive species is a carbene
and not a free radical as was first expected. There have also been several
attempts to model the kinetics of the reaction and the most successful
attempt has been made by Baughman.[31] The conversion/time curves for
the polymerization of many diacetylenes have characteristic sigmoidal

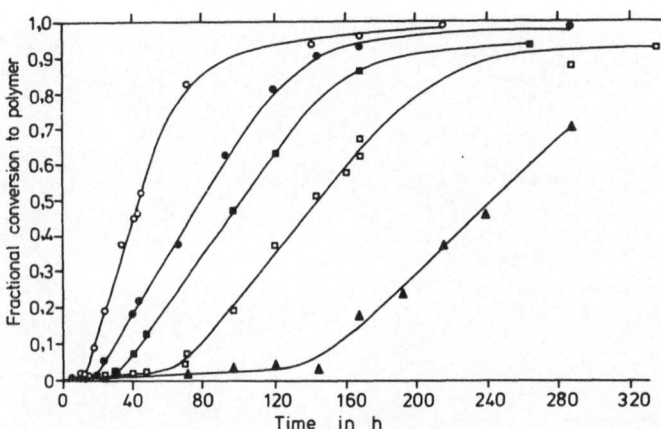

FIG. 4. Time–conversion curve for the isothermal solid-state polymerization of EUHD: ▲—70°C; □—75°C; ■—78°C; ●—80°C; ○—86°C.(After Reference 27.)

shapes as shown in Fig. 4 for the thermal polymerization of EUHD. The polymerization of most diacetylenes involves a small (~5%) change in crystal lattice parameters, and this leads to internal strains as the polymer chains form inside the monomer lattice. This behaviour and the shape of the curves has been explained successfully in terms of crystal strain models although the fact that it also works for polymerizing polydiacetylene systems which are not single-phase raises some doubt about the generality of the model and also suggests that some of the reported fits[27] may be fortuitous.

Since the high modulus fibre-forming diacetylenes are insoluble once polymerized, it is impossible to determine the molar mass or degree of polymerization of the polymers. The fibres behave as if the molar mass is effectively infinite, although recent work on soluble diacetylene derivatives with long aliphatic side-groups[32] has suggested that solid-state polymerization leads to degrees of polymerization of the order of 1000–1500. It is debatable, however, as to how relevant this result is to the high modulus derivatives with short side-groups.

## 3. STRUCTURE

### 3.1. Crystal Morphology

Polydiacetylene single crystals can be obtained essentially in two crystal forms, either as lozenges or as fibres. The morphology is controlled by the

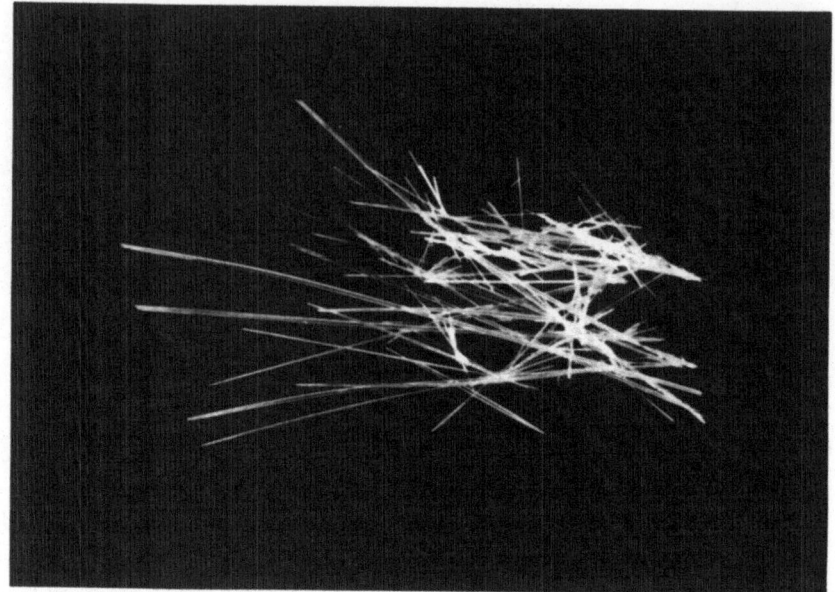

FIG. 5.   Single crystal of EUHD monomer crystallized from distilled water. (After Reference 29 by permission of the publishers, John Wiley & Sons.)

conditions under which the monomer is crystallized from solution, although the exact reasons why a particular morphology is obtained are not really understood. TSHD (Table 1) is normally only found in the form of lozenges when crystallized from most solvents,[13] whereas DCHD is usually obtained as fibres[25] the aspect ratio of which depends upon the solvent, solution concentration and crystallization temperature. In contrast, EUHD can be obtained in three crystal forms. The form which can undergo solid-state polymerization to give fully polymerized single crystal fibres[27] is shown in Fig. 5. Good polymerizable fibres of PUHD (Table 1) are obtained when it is crystallized from dioxane/water mixtures and dioxane molecules are trapped interstitially in the structure.[33] In each of these cases the morphology is retained when the monomer crystals are transformed into polymer by the solid-state polymerization reaction.

A particularly convenient way of investigating the structure of polydiacetylenes is by transmission electron microscopy. Monomer crystals which are sufficiently thin ($\sim 100$ nm) to allow penetration of the electron beam can be readily produced by allowing a droplet of dilute monomer solution to evaporate on a carbon support film on an electron

TABLE 2

UNIT CELL PARAMETERS FOR THE POLYMERS OF THE FOUR DIACETYLENE DERIVATIVES
IN TABLE 1; THE CHAIN DIRECTION IS $b$

| Derivative | Unit cell parameters | | | | Ref. |
|---|---|---|---|---|---|
| | $a$ (nm) | $b$ (nm) | $c$ (nm) | $\beta$ (°) | |
| TSHD | 1·449 | 0·491 | 1·494 | 118 | 35 |
| PUHD[a] | 0·489 | 1·253 | 1·677 | 96·8 | 36 |
| DCHD | 1·287 | 0·491 | 1·740 | 108·3 | 37 |
| EUHD | 1·796 | 0·488 | 1·514 | 98·4 | 27 |

[a] Triclinic ($P\bar{1}$) with $\alpha = 69\cdot3°$, $\gamma = 96\cdot2°$, and $a$ as the chain direction.

microscope grid.[34] The monomer can then be polymerized by heating or most conveniently by exposure to the electron beam in the microscope.

## 3.2. Crystal Structure

The crystal structures of several polydiacetylenes have been determined to a high degree of accuracy using X-ray diffraction methods.[35-37] This can be contrasted with conventional polymers for which the crystal structures[38] determined using oriented polycrystalline samples are much less accurately known. The unique single crystal nature of polydiacetylenes has enabled the positions of the atoms in the unit cells and bond angles to be measured to a high degree of precision. Many diacetylene monomers and polymers have monoclinic $P2_1/c$ crystal structures (indexing the chain direction unconventionally as $b$); the unit cell parameters of several common monomers and polymers are listed in Table 2. This detailed knowledge of the crystal structures is invaluable for the correlation of physical properties with structure in these materials.

## 3.3. Electron Microscopy

Although X-ray diffraction is by far the most accurate method of determining crystal structures, considerable extra information can be obtained using electron microscopy with the ability of obtaining high magnification images of the structure and performing simultaneous electron diffraction. Figure 6(a) shows a micrograph of single crystal fibres of polyDCHD along with a diffraction pattern obtained from one of the crystals. The diffraction pattern confirms that the chain direction is parallel to the fibre axis. In addition the pattern shows that the polymer fibres are true single crystals with no amorphous scattering.

One problem normally encountered with the study of organic materials

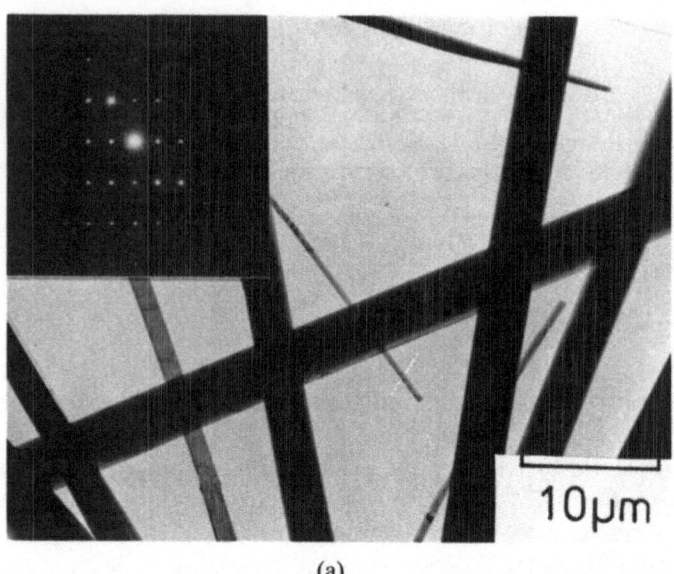

(a)

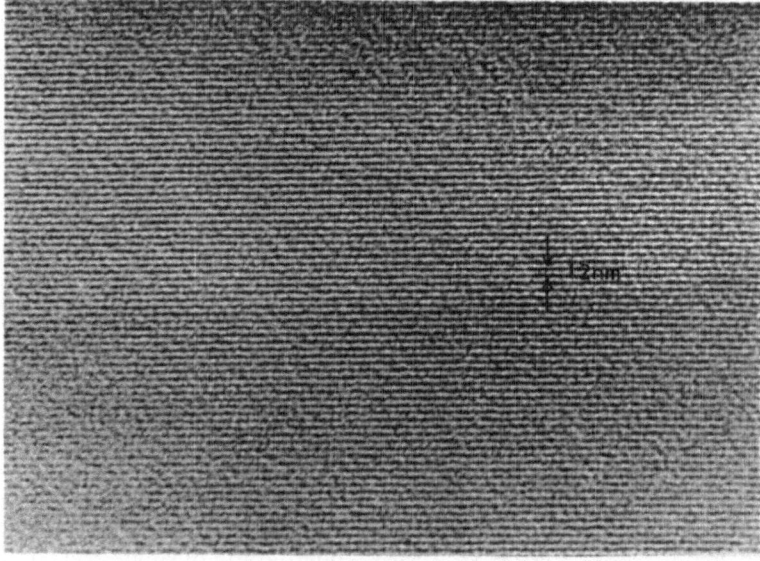

(b)

FIG. 6.   Transmission electron micrographs of fibrous crystals of polyDCHD. (a) Low magnification micrograph showing several crystals on a carbon support film with diffraction pattern (inset). (b) High magnification micrograph showing lattice planes with a spacing of 1·2 nm parallel to the chain direction. (After Reference 26 by permission of the publishers, John Wiley & Sons.)

in the electron microscope is that they are prone to radiation damage whereby the crystal structure is destroyed by exposure to the electron beam.[39] It has been found that most polydiacetylenes are relatively stable but that polyDCHD is particularly outstanding in its ability to resist damage in the electron beam, being at least 20 times more stable than polyethylene[40] which is the most widely studied crystalline polymer. This high stability has allowed detailed investigations to be made of the structure of polyDCHD at high magnification. In particular it has been possible to image the crystal lattice directly in the microscope[40,41] as is shown in Fig. 6(b) where lattice planes parallel to the chain direction are imaged. All these observations indicate that there is perfect molecular alignment and a high degree of crystal perfection in the fibres. This can be contrasted with the highly defective polycrystalline structures found by similar electron microscope observations on polymer fibres produced using techniques such as mechanical orientation[42] and solvent spinning of liquid crystalline solutions of rigid-rod polymers.[43] This difference in structure is reflected directly in the differences in the mechanical properties between the single crystal fibres and fibres produced using other techniques, as will be shown later.

One factor which controls the Young's modulus of polymer crystals is the sideways packing of the molecules, and this can be elucidated by viewing the crystals parallel to the chain direction. Recent work[44,45] has shown that it is possible to prepare lamellar crystals of polyDCHD (Fig. 7(a)) in which the molecules are perpendicular to the crystal surface and hence parallel to the electron beam in the microscope. Micrographs obtained at high magnification show a full structural image which, using conventional techniques, can be enhanced to remove any noise.[46] Figure 7(b) shows an enhanced molecular image of a polyDCHD crystal viewed parallel to the molecules such that the projections of the side-groups overlap so that the profiles of the molecules can be seen. Also, in the inset, there is a computer-simulated image of the structure obtained from knowledge of the crystal structure[37] of polyDCHD. It can be seen that the molecules consist of a small backbone with two relatively large dumb-bell shaped side-groups. It will be shown later that Young's moduli of polydiacetylene fibres are controlled directly by the size of these side-groups.

### 3.4. Defects
The presence of defects is known to affect the mechanical behaviour of crystals of most materials, and polymer single crystals are no exception.

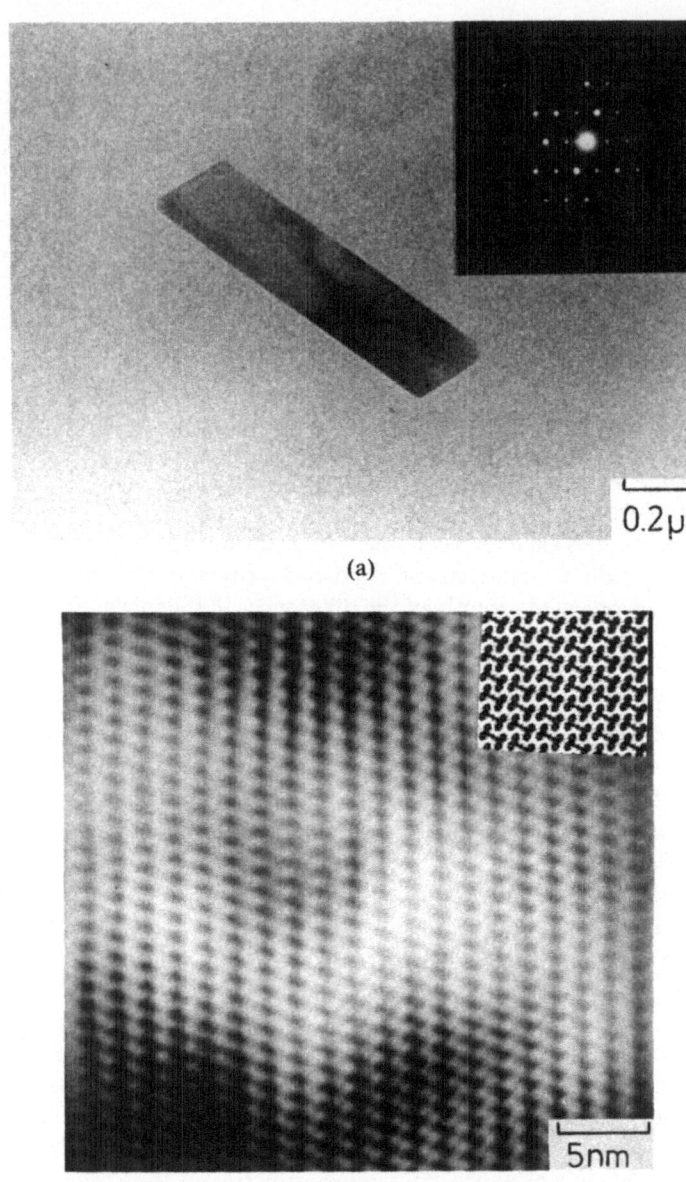

0.2 μm

(a)

5nm

(b)

FIG. 7. Transmission electron micrographs of lamellar crystals of polyDCHD. (a) Low magnification micrograph showing a crystal on a carbon support film with electron diffraction pattern (inset). (b) Electronically filtered high-resolution micrograph showing individual molecules projected parallel to the chain direction with simulated image inset. (After Reference 44 by permission of the publishers, Chapman and Hall.)

The study of defects in metal crystals has taken place over the past 20 years and there is now a high level of understanding of the structure of defects in metal crystals and their influence upon properties. The chain-folded solution-grown lamellar single crystals of conventional polymers prepared from dilute solution have allowed investigations to be carried out on certain types of dislocations.[47-49] Unfortunately, the lamellar morphology of these crystals limits the viewing direction in the microscope to parallel to the chain direction, so defects which can only be seen by viewing in other directions cannot be investigated. It is therefore only possible to study a restricted number of defects. Also, radiation damage[39,40] restricts the time available to analyse defects using different microscope conditions. The advent of polydiacetylene single crystals has allowed detailed studies of several types of defects found in polymer single crystals to be carried out because of their better stability in the microscope, their well defined structure and the ability to view crystals at different angles to the chain direction.

Most of the detailed studies on defects in polydiacetylene single crystals have been carried out on polyTSHD. Edge dislocations with Burgers vectors parallel to the chain direction have been reported in crystals of this material.[50,51] The density of these dislocations is found to be of the order of $10^{13} \, m^{-2}$, and in some cases they can line up to form a small-angle grain boundary.[50] Edge dislocations with Burgers vectors perpendicular to the chain direction consisting of an extra half-plane of chain-ends have been imaged in polyDCHD using high resolution techniques.[40] It is likely that both of these types of dislocations are present initially in the monomer crystals and become frozen into the polymer during polymerization. It is thought that although both types of dislocations are probably capable of moving in the monomer they cannot move in the polymer without breaking covalent bonds and so are probably not involved in deformation. Stacking faults have also been seen in crystals of polyTSHD,[52] but since the planes of the faults are parallel to the chain direction they could only be involved in the deformation of crystals stretched or compressed in directions perpendicular to their molecular axes.

Surface defects such as steps are known to have an important effect upon the fracture behaviour of polydiacetylene single crystal fibres, so the geometry of such defects has been examined in detail.[26] Examples of steps on the surfaces of polyDCHD fibres are given in Fig. 8. As with the dislocations, it is likely that the steps form during growth of the monomer crystals from solution. Indeed, in the microscopic fibres (Fig. 8(b)) it can be

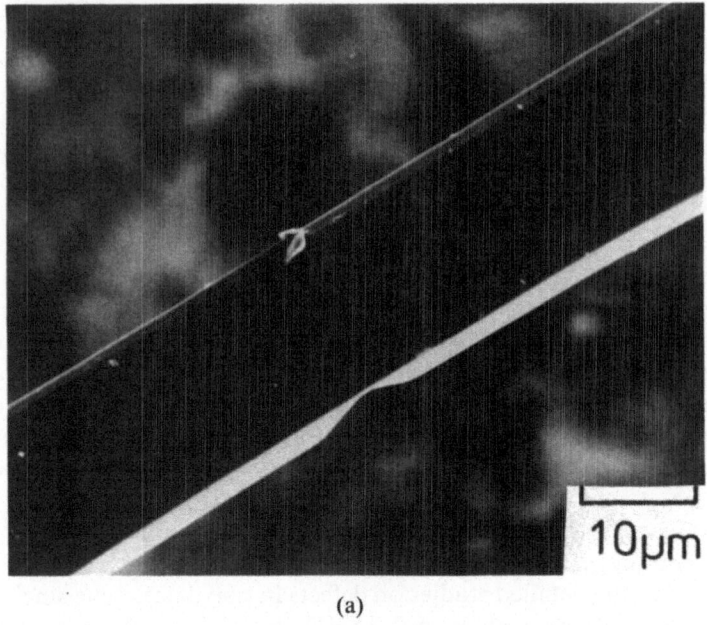

(a)

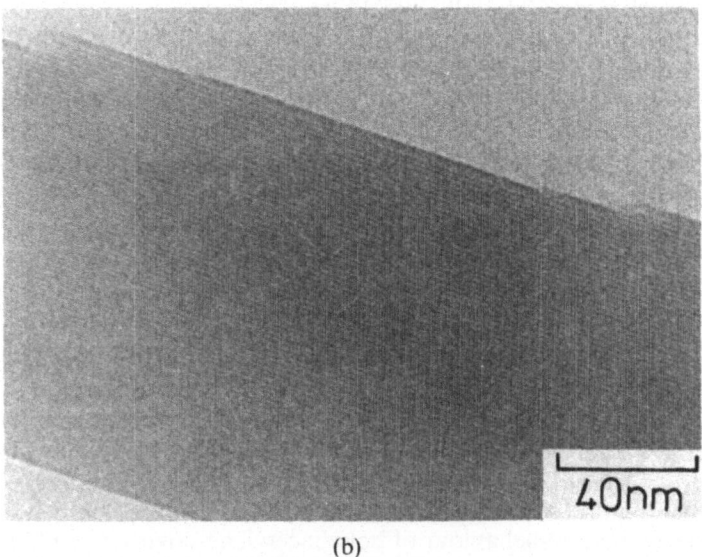

(b)

FIG. 8. Surface steps on fibrous crystals of polyDCHD. (a) Scanning electron micrograph showing large step. (After Reference 26 by permission of the publishers, John Wiley & Sons.) (b) High resolution transmission electron micrograph showing lattice image with molecular steps. (Courtesy P. H. J. Yeung.)

seen that crystal growth has taken place by the deposition of individual molecular layers. Although the two fibres have very different diameters, the size of the steps in both cases is of the order of 1/10 to 1/20 of the fibre diameter, and it will be shown that this has an important effect upon the size dependence of the strengths of the crystals (Section 6.2).

# 4. ELASTIC DEFORMATION

## 4.1. Stress/Strain Behaviour

The possibility of growing single crystals of polydiacetylenes with centimetre dimensions has enabled the stress/strain behaviour of polymer single crystals to be determined for the first time using conventional mechanical testing methods. In 1974 Baugham et al.[33] demonstrated that fibre-like single crystals of polyPUHD (Table 1) could be deformed elastically to strains of over 3%. The crystals were found to have values of Young's modulus in the chain direction of the order 45 GPa. Such high degrees of stiffness have also been found for other polydiacetylenes with Young's modulus values of 45 GPa being determined for polyDCHD[26] and 62 GPa for polyEUHD.[29]

A typical stress/strain curve for a polyDCHD single crystal fibre is given in Fig. 9(a). The curve is linear up to a strain of about 1·8% and there is a slight decrease in slope above this strain until fracture occurs at a strain of about 2·8%. Loading and unloading take place along the same path,[26] indicating a lack of hysteresis in the deformation. The fracture strain is found to depend upon the fibre diameter, decreasing as the diameter is increased.[26]

The stress/strain curve for a polymer single crystal in Fig. 9(a) gives a clear indication of how deformation takes place on the molecular level, as it corresponds essentially to a molecular stress/strain curve. Baughman et al.[33] suggested that the slight deviation from Hooke's law above about 2% strain, which is clearly not a yield process, might be due to the anharmonic part of the interaction potential between neighbouring atoms on the polymer chain. Since the polydiacetylene fibres are highly perfect polymer single crystals the deformation directly involves the stretching and bending of bonds along the polymer backbone. This has been confirmed using resonance Raman spectroscopy[28,53,54] where it has been shown that the frequencies of the C—C, C=C and C≡C stretching modes in polydiacetylenes depend upon the deformation of the crystals, decreasing with applied strain. An example of the variation of the frequency of the

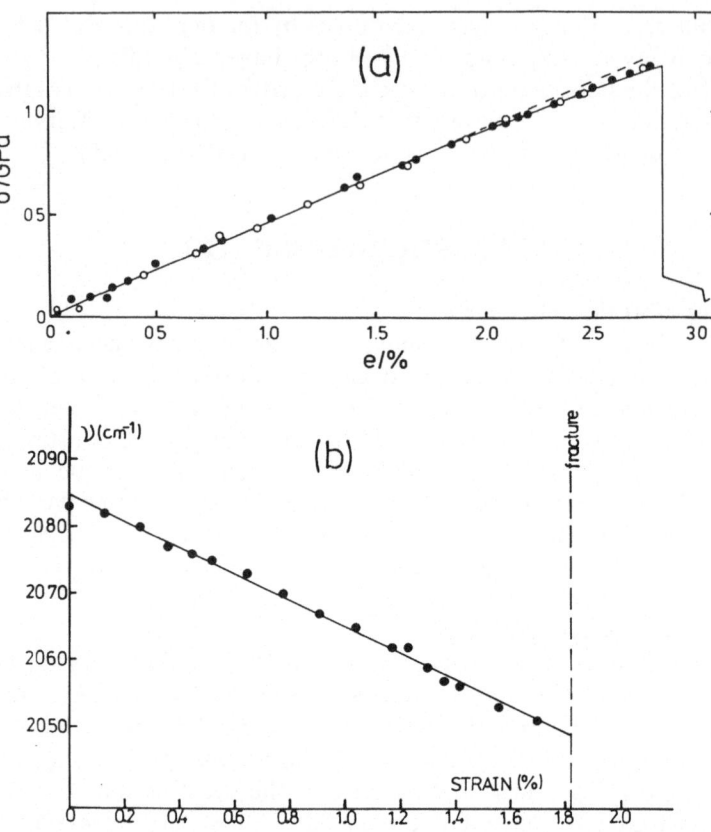

FIG. 9.   Elastic deformation of polyDCHD fibres. (a) Stress/strain curve. (After Reference 28 by permission of the publishers, John Wiley & Sons.) (b) Change in C≡C bond stretching frequency with applied strain. (After Reference 54.)

C≡C stretching mode for polyDCHD as a function of applied strain is shown in Fig. 9(b). The consequent reduction in force constants is one of the factors leading to the reduction in the slope of the stress/strain curves at high strains.[26,29,33]

## 4.2. Factors Controlling Young's Modulus

Baughman and coworkers[33] pointed out that the levels of chain direction modulus displayed by polydiacetylene single crystals were extremely high when account was taken of the high cross-sectional area, $A$, of the molecular chains due to the relatively large side-groups on the polydiacetylene molecules. This is demonstrated in Fig. 10(a) where the chain direction modulus, $E$, of various polydiacetylenes is plotted against the reciprocal of the area supported by each chain in the crystal, $1/A$ (determined from knowledge of the crystal structures, Table 2). It can be seen that the data fall close to a straight line demonstrating clearly the direct relationship between the modulus and crystal structure. The value of $1/A$ for polyethylene is also indicated in Fig. 10(a). Assuming that the two backbones have about the same stiffness, a modulus of the order of 250 GPa is predicted for polyethylene, which is in line with current theoretical predictions.[55] The realization that this per-chain modulus is nearly as high as that of diamond[56] has led to strenuous efforts being made to align molecules in polyethylene fibres.[1-7] In the case of polydiacetylenes the alignment is a natural consequence of their production via topochemical solid-state polymerization. Hence the way forward with these materials is in the chemical synthesis of monomers with smaller substituent R groups (Table 1) that will form fibres and polymerize fully to give perfect polymer crystals.

It is also found that for a particular diacetylene derivative the Young's modulus depends upon degree of conversion from monomer to polymer and the method of polymerization. This is demonstrated in Fig. 10(b) where the dependence of $E$ upon the conversion of EUHD to polymer using either heat or $\gamma$-rays is shown.[29] It can be seen that there is a linear increase in the modulus with the volume fraction of polymer in the crystals for thermal polymerization whereas the moduli of the crystals produced by $\gamma$-ray polymerization are somewhat lower. Although it is known that the thermally polymerized crystals are crystallographically less perfect, it is thought that the high doses of radiation needed to polymerize EUHD damage the crystals and so reduce their moduli.

## 4.3. Theoretical and Measured Moduli
Although it is the Young's modulus in the chain direction that is of the greatest importance for engineering applications, it must be remembered that a large number of elastic constants are needed to describe fully the elastic behaviour of a crystal.[57] Polydiacetylene single crystals usually possess monoclinic symmetry and therefore have 13 elastic constants. There have been several attempts to calculate the 9 elastic constants for orthorhombic polyethylene[58] but as yet there have been no similar calculations for a polydiacetylene. The calculation ought to be feasible as the crystal structures and atomic positions of several polydiacetylenes are known to a high degree of accuracy.[35-37] However, there are also a large

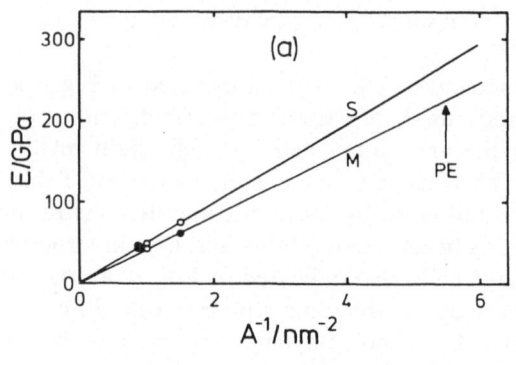

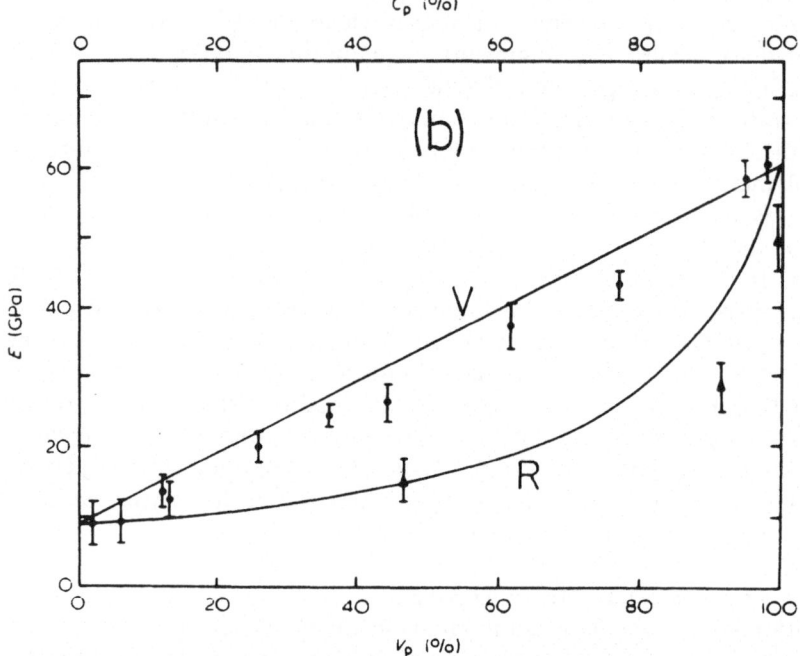

FIG. 10. (a) Dependence of the Young's modulus of polydiacetylene single crystal
fibres upon the reciprocal of the area supported by each polymer chain. The S line
and the open circles are for values calculated using force constants determined by
Raman spectroscopy. The M line and the closed circles are for mechanically
measured ones. The arrow corresponds to the value of $1/A$ for polyethylene. (After
Reference 26 by permission of the publishers, John Wiley & Sons.) (b) Variation of
Young's modulus with conversion to polymer ($C_p$) and volume fraction of polymer
($V_p$). The circles are for thermally polymerized fibres and the triangles for $\gamma$-ray
polymerized ones. (After Reference 29 by permission of the publishers, John Wiley
& Sons.)

number of different bonds in each unit cell, making such a calculation extremely complex.

The modulus of polyTSHD crystals deformed in the chain direction has been calculated by Batchelder and Bloor[53] employing the method of Treloar.[59] They used force constants measured from Raman spectroscopy and estimated Young's modulus for polyTSHD crystals in the chain direction to be 50 GPa which is in good agreement with the value of 45 GPa determined for polyDCHD[26] which has similar unit cell dimensions. The method of Treloar is relatively simple but it cannot be extended to determine the full set of elastic constants since it only takes into account the covalent bonding along the polymer backbone.[59]

It has been possible to measure some of the 13 elastic constants for polydiacetylene single crystals using Brillouin scattering[60,61] and sound velocity measurements.[62] Wegner and coworkers[60] employed Brillouin scattering to determine 6 of the elastic constants of the monomer and 3 for the polymer of TSHD at room temperature. Rehwald *et al.*[62] showed that sound velocity measurements enabled 9 of the elastic constants to be determined for TSHD monomers and polymers over a wide range of temperature, although the errors involved in some of the determinations were rather large. Values of the elastic constants $c_{ik}$,[57] determined for TSHD monomer and polymer, are compared in Table 3. It can be seen that, on the whole, the agreement between the elastic constants determined by the different techniques is relatively good. In particular, the values of $c_{22}$ for the polymer (which reflects the deformation in the chain direction)

TABLE 3

ELASTIC CONSTANTS $c_{ik}$ DETERMINED AT ROOM TEMPERATURE FOR TSHD MONOMER AND POLYMER USING BRILLOUIN SCATTERING[60] AND SOUND VELOCITY MEASUREMENTS[62]

| Elastic constants $c_{ik}$ (GPa) | Monomer | | Polymer | |
|---|---|---|---|---|
| | Ref. 62 | Ref. 60 | Ref. 62 | Ref. 60 |
| $c_{11}$ | 5·5 | 8·7 | ~7 | 12·9 |
| $c_{22}$ | 6·7 | 9·7 | 43·3 | 43·7 |
| $c_{33}$ | 3·5 | 8·2 | <9·2 | — |
| $c_{44}$ | 1·0 | — | 1·07 | — |
| $c_{55}$ | 3 | 3·4 | ~3 | 6·9 |
| $c_{66}$ | 0·9 | 1·5 | 1·03 | — |
| $|c_{15}|$ | ~1 | — | <3·1 | — |
| $|c_{35}|$ | 1·0 | — | — | — |
| $|c_{46}|$ | 0·05 | — | ~0·03 | — |

are in close agreement. In addition, it can be seen that, while most of the values of $c_{ik}$ remain mainly unaffected by the solid-state polymerization reaction, there is a large increase in the value of $c_{22}$ by a factor of about 6. This reflects the replacement of weak van der Waals bonding with strong covalent bonds along the polymer backbone. It is similar to the increase in Young's modulus with conversion shown in Fig. 10(b) for polyEUHD, and again emphasizes the high stiffness that can be achieved from polymer molecules.[56]

## 5.  PLASTIC DEFORMATION

Because of the restrictions of not breaking covalent bonds during deformation, polymer crystals are only capable of undergoing a limited amount of plastic deformation.[63] In polycrystalline samples of conventional polymers much of the deformation is taken up by the non-crystalline amorphous phase.[63] Since there is no amorphous material in single crystals of polymers such as polydiacetylenes, it is found that their ability to undergo plastic deformation is severely restricted, so they are highly resistant to creep.[26,29] However, polydiacetylenes are found to be capable of undergoing some limited plastic deformation through twinning.[64,65]

### 5.1. Creep
One of the most striking aspects of the mechanical properties of polydiacetylene single crystals is that it is not possible to measure any time-dependent deformation (or creep) when crystals are deformed in tension parallel to the chain direction.[26,29] This behaviour is demonstrated in Fig. 11 for a polyDCHD single crystal held at constant stress at room temperature, and preliminary measurements have indicated that creep could not be detected during deformation at temperatures up to at least 100°C.[26]

Creep and time-dependent deformation are normally a serious problem in the use of high-modulus polymer fibres such as polyethylene[66] in engineering applications. Such oriented fibres produced by drawing or spinning contain a high density of defects such as chain-ends, loops and entanglements. These allow the translation of molecules parallel to the chain direction during deformation, which leads to creep. However, recent work[67] has shown that the creep in such fibres can be dramatically reduced by radiation cross-linking. In contrast, since polydiacetylene

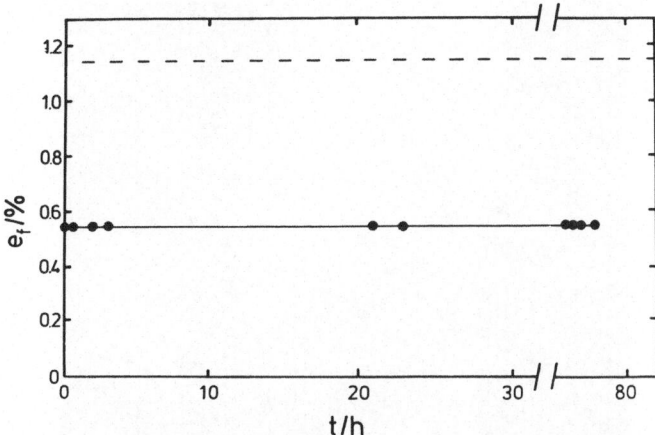

FIG. 11. Variation of fibre strain with time for a polyDCHD fibre held at constant stress corresponding to 50% of its fracture strain, given by the dashed line. (After Reference 26 by permission of the publishers, John Wiley & Sons.)

single crystal fibres contain perfectly aligned long polymer molecules (cf. Fig. 6(b)), there is no mechanism whereby creep can take place even at high temperatures. In addition, polydiacetylenes tend to degrade rather than melt, and this should be contrasted with polyethylene which melts[57] at about 140°C and so has only very limited high-temperature applications.

## 5.2. Twinning

Studies on the deformation of polydiacetylene single crystals[64,65] have demonstrated that polymer single crystals twin when deformed in compression parallel to the chain direction. This type of deformation leads to the formation of twins which involve the molecules kinking over at a well defined angle.[64,65] Figure 12 shows a twin in a polyDCHD fibre[68] along with the corresponding diagram of the molecular displacements involved. The process can be differentiated from the formation of kink-bands which are found in other high modulus fibres[69] since the material within the twinned region has the same crystal structure as the undeformed crystal. There is also a mirror image orientation relationship between the deformed and undeformed regions,[64] which is not the case with kink bands.[69]

Until this work on the polydiacetylene single crystals it was thought that twinning in polymer crystals would not be able to take place by the bending of the molecular chains.[70] However, in 1976 Pietralla[71] postulated that this type of twinning might occur in semicrystalline

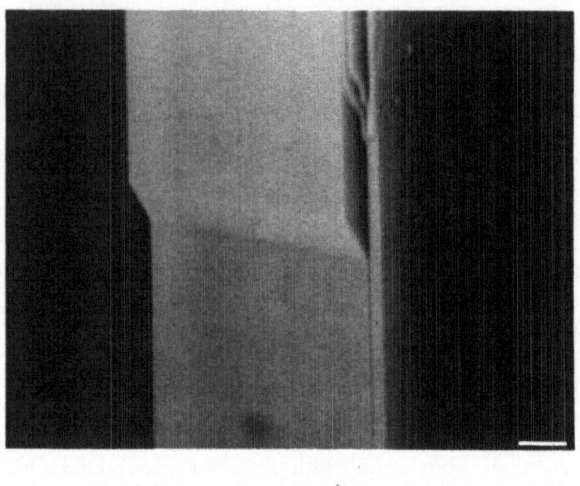

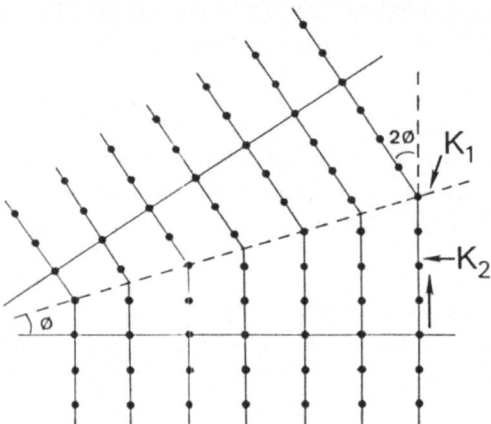

FIG. 12. Scanning electron micrograph of a twin in a polyDCHD fibre, along with a molecular sketch of the twin. (After Reference 68 by permission of the publishers, Chapman and Hall.)

polymers, and Bevis[72] subsequently performed a detailed analysis of this 'chain twinning' for the case of polyethylene crystals. Nevertheless, the investigations into deformed crystals of polydiacetylenes[64,65] have provided the first direct evidence for chain twinning taking place. However, a survey of some of the old literature on chain-extended polyethylene lamellae[73] has demonstrated that the twins are probably also present in such structures.

The ability of polymer single crystal fibres to undergo chain twinning

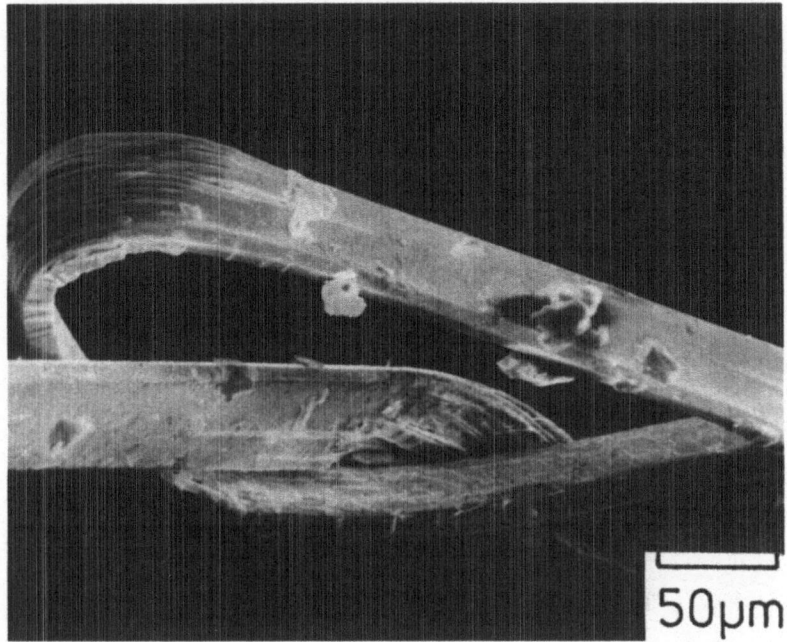

50μm

FIG. 13.   Scanning electron micrograph of a knotted polyDCHD fibre. (After
Reference 26 by permission of the publishers, John Wiley & Sons.)

has important consequences for their use in composites. For example, it is
possible to tie knots in polyDCHD fibres;[26] an example of a knotted
crystal is shown in Fig. 13. This process involves a high degree of
deformation but the crystals are able to cope with the high strains by
undergoing chain twinning on the inside surface which is subject to
compression in the chain direction. There tends to be some cracking
parallel to the chain direction, but the crystals remain relatively intact even
when pulled into a tight knot.[26] This behaviour has implications for the
use of such fibres in composites, as their ability to absorb strain by
twinning means that the fibres should not break up and undergo 'fibre
attrition' as easily as other fibres such as those of glass and carbon. It has
also been found recently that the polydiacetylene single crystal fibres tend
to twin during fabrication in composites when thermosetting matrices
such as epoxy resins are used.[68] Resin shrinkage during both curing and
cooling from the cure temperature imposes a compressive stress parallel to
the fibre axis which can twin the crystals. The consequences of this
behaviour are discussed further in Section 7.2.

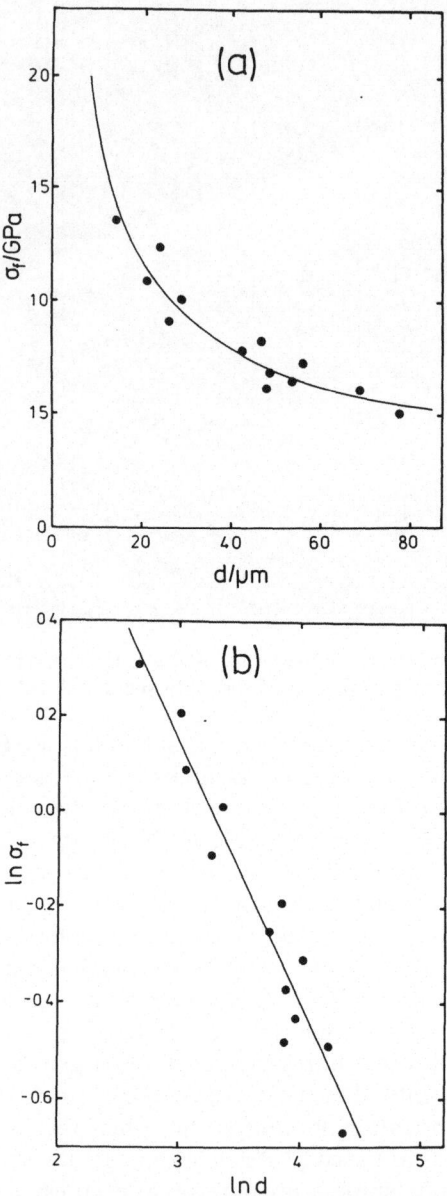

FIG. 14.   Dependence of the fracture strength of polyDCHD fibres upon the fibre diameter: (a) linear plot; (b) log–log plot. (After Reference 26 by permission of the publishers, John Wiley & Sons.)

## 6. FRACTURE

Polymer single crystals readily undergo cleavage parallel to the chain direction, reflecting the relative strength of covalent bonding compared with the van der Waals bonding between the polymer molecules. In polydiacetylenes cleavage takes place preferentially on certain crystallographic planes, and in polyTSHD crystals a correlation has been found between the presence of stacking faults[52] and easy cleavage on (102) planes.[74] However, the most interesting aspect of the fracture behaviour of polymer single crystal fibres is the high strength they can exhibit when deformed parallel to the chain direction; this is described in detail below.

### 6.1. Fracture Strength

Investigations into the fracture behaviour of polydiacetylene single crystal fibres have revealed a strong dependence of the fracture stress, $\sigma_f$, upon the fibre diameter, $d$.[26,29,33] This is demonstrated for polyDCHD fibres in Fig. 14(a) and it is similar to the size dependence reported earlier for inorganic high-strength fibres[75,76] where the dependence of $\sigma_f$ upon $d$ was found to follow a relation of the form

$$\sigma_f \propto 1/d \tag{1}$$

This behaviour was thought to be due to the presence of surface defects which give rise to a stress concentration when the fibres are deformed. Since the size of the defects in the inorganic fibres was found to scale with the fibre diameter, the size dependence could be predicted.[77] However, more detailed examination of the data in Fig. 14(a) has shown that eqn (1) is not accurately obeyed for polydiacetylene single crystal fibres. In addition, theoretical calculations have shown that a different relationship is expected.[26,29] The data in Fig. 14(a) have been re-plotted in Fig. 14(b) in the form of a log/log plot, and it can be seen that the dependence of $\sigma_f$ upon $d$ is given more accurately by

$$\sigma_f \propto 1/d^{0.55} \tag{2}$$

This is precisely the dependence predicted from the theoretical considerations outlined below.

### 6.2. Theoretical Strength

The magnitude of the stress concentration will depend upon the geometry of the surface defects. Examples of defects found on the surfaces of polyDCHD crystals were shown in Fig. 8. Although the crystals are of very different sizes, it was pointed out that the defects were roughly 1/10 to 1/20 of the crystal diameters. It can be shown[76] that, in general, the stress concentration factor, $\eta$, at the root of a step or notch is given by

$$\eta = \sigma_0/\sigma_a = k_1 + k_2 a^{1/2}\rho \qquad (3)$$

where $\sigma_a$ is the applied stress, $\sigma_0$ the stress at the root of the defect, $a$ the notch or step depth, $\rho$ the radius of curvature at the root of the defect, and $k_1$ and $k_2$ are constants which depend upon the geometry of the defects.[76,77]

It can be seen from Fig. 8, and it has been shown elsewhere,[29] that the parameter $a$ scales with the fibre diameter such that

$$a \approx d/m \qquad (4)$$

where $m$ is a constant which is of the order of 10–20. Putting eqn (4) into eqn (3) and setting $\sigma_a = \sigma_f$ at fracture and rearranging gives

$$\frac{1}{\sigma_f} = \frac{k_1}{\sigma_0} + \frac{k_2}{\sigma_0(\rho m)^{1/2}}d^{1/2} \qquad (5)$$

where $\sigma_0$ now becomes the strength of a defect-free crystal (i.e. the 'theoretical strength'). Hence it is predicted that a plot of $1/\sigma_f$ versus $d^{1/2}$ should give a straight line of slope $k_2\sigma_0(\rho m)^{1/2}$ with an intercept of $k_1/\sigma_0$. The data from Fig. 14(a) have been plotted according to eqn (5) in Fig. 15. It can be seen that the data fall on a straight line and give an intercept close to the origin. In addition, knowledge of the detailed geometry of the surface defects allows the value of the theoretical strength, $\sigma_0$, to be

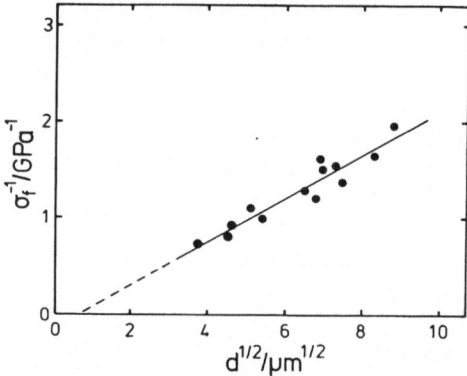

FIG. 15. Variation of the reciprocal of the fibre strength with square root of fibre diameter for polyDCHD fibres. (After Reference 26 by permission of the publishers, John Wiley & Sons.)

determined from the slope of the line in Fig. 15. The theoretical strength of polyDCHD crystals has been determined to be $3 \pm 1$ GPa,[26] the rather large error being due to the uncertainty in determining the defect geometry.

### 6.3. Molecular Fracture

Frank[56] pointed out several years ago that polymer molecules can have very high values of strength when deformed parallel to the direction of their molecular chains. The determination of the theoretical strength of polydiacetylene single crystal fibres allows the strength of individual molecules to be estimated. From knowledge of the crystal structure of polyDCHD (Table 2),[37] it can be shown that each molecule supports a cross-sectional area, $A$, of about 1 nm$^2$. The theoretical strength of 3 GPa therefore corresponds to a force to break molecules of about 3 nN and a fracture strain of the order of 6–8%. It is of interest to compare this with the theoretically calculated values of strengths of covalently bonded polymer molecules.[76,78,79] Kelly[76] has estimated the strength of a polyethylene molecule as about 6 nN but this is thought to be rather high. Kausch[79] has shown that a covalently bonded polymer molecule should be broken by a force of about 3 nN, which is identical to the value determined for polyDCHD. A molecular strength of this magnitude corresponds to a fracture stress of the order of 20 GPa for polyethylene single crystals where the area supported by each molecule is considerably smaller than for polyDCHD. However, it has not yet been possible to make single crystals of polyethylene with macroscopic dimensions, so even highly oriented polyethylene fibres are found to have strength values only of the order of 4 GPa,[3,5] which is well below the theoretical strength.

## 7. COMPOSITES

It is well established[80] that composites produced by incorporating high-modulus fibres in a brittle matrix such as an epoxy resin can have outstanding mechanical properties. Recent examples have included composites produced with high-modulus polyethylene fibres[81] and aromatic polyamide fibres.[82] Polydiacetylene single crystal fibres also offer considerable promise as reinforcing fibres in polymeric matrices because they have the following properties which have been described in previous sections: (i) high stiffness; (ii) high strength; (iii) low creep; (iv) good thermal stability; (v) low density.

Recent investigations[68,83,84] into the behaviour of polydiacetylene single crystal fibres in epoxy resin matrices have shown not only that such composites have promising mechanical properties but that important fundamental details of the mechanisms of fibre reinforcement can also be revealed from their study.

### 7.1. Micromechanics of Reinforcement

It was pointed out in Section 4.2 that the vibration frequencies of certain main-chain Raman active modes were found to change with the level of applied strain.[28,53,54] In particular it was found that the $C \equiv C$ triple bond stretching frequency changes by the order of $20 \, cm^{-1}$ for 1% of strain,[53] as is shown in Fig. 9(b). This property can be used to determine the strain in a polydiacetylene fibre subjected to any general state of stress. The strain can be measured to a high degree of spatial resolution and accuracy as beam diameters as small as $\sim 10 \, \mu m$ can be used and changes in frequency can be determined to $\sim 1 \, cm^{-1}$. The fibre therefore behaves as though it has an internal molecular strain gauge, and the Raman technique has recently been employed to monitor the point-to-point variation in strain in polydiacetylene fibres in epoxy composites.[68,83,84]

The first case investigated was that of a single short fibre in a matrix subjected to an overall strain, which is a classical problem in the theory of fibre reinforcement.[80,85] Model specimens were fabricated and an example is shown in Fig. 16(a). When the specimen is deformed the matrix strain is measured using a strain gauge adhered to the surface of the specimen, and the point-to-point variation in strain is monitored using Raman spectroscopy. Typical results[83] are shown in Fig. 16(b) where the fibre strain is plotted as a function of position along the fibre for different levels of applied matrix strain; it was found that the results agreed moderately well with theoretical predictions of Cox.[85] For example, it can be seen that at higher levels of matrix strain the fibre strain rises from the end to a constant value along the length of the fibre and then falls off at the other end as predicted by the theory. These regions of rise and fall of strain are known as the 'critical length'[80] and it was also demonstrated[83] that the critical length was proportional to the fibre diameter. Although this result was expected from the theoretical analyses,[83,85] this is the first time that it has been demonstrated by direct experiment for any composite system.

### 7.2. Resin Shrinkage

A more detailed analysis of the data in Fig. 16(b) shows important deviations from theoretical predictions. For example, at low levels of

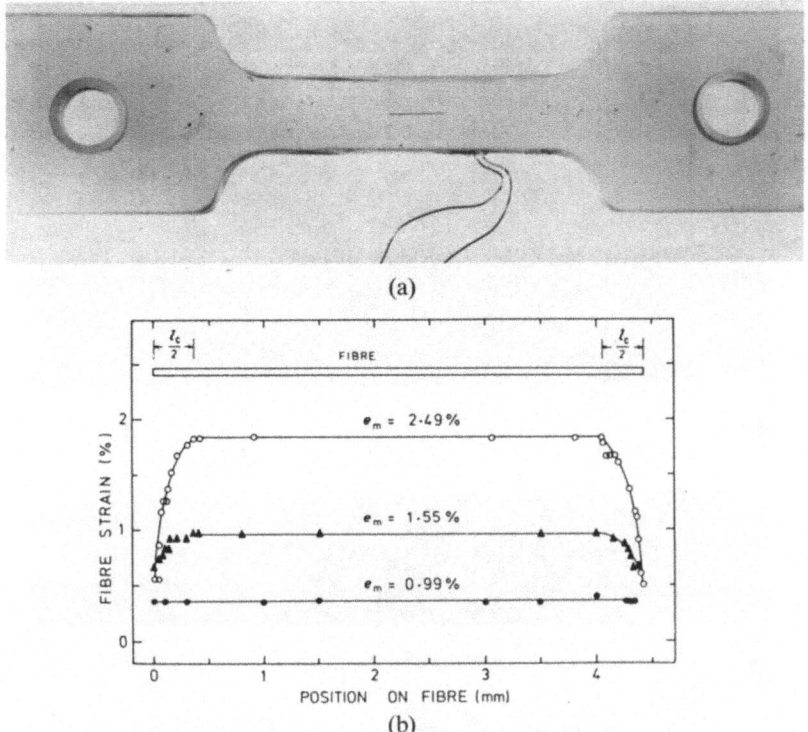

(a)

(b)

FIG. 16.   PolyDCHD/epoxy resin single fibre composites. (a) Specimen with strain gauge attached. (b) Variation of fibre strain with position along the fibre for different levels of matrix strain, $e_m$. (After Reference 83 by permission of the publishers, Chapman and Hall.)

applied matrix strain the fibre strain is small and does not vary with position along the fibre. Also the fibre strains and matrix strains are not equal in the middle of the fibres as would be expected from theoretical considerations.[80,85] Although it was originally thought[83] that this was due to inappropriate assumptions in the theory and viscoelastic deformation in the matrix, it has recently been demonstrated that the behaviour is due to resin shrinkage during both curing and cooling from the cure temperature.[68] Since the thermal expansion coefficients of the single crystal fibres in the chain direction are extremely small or even negative,[86] the resin shrinkage imposes compressive stresses on the fibres parallel to their axes. These compressive stresses then give rise to twinning in the fibres as is shown in Fig. 17. Tensile deformation parallel to the

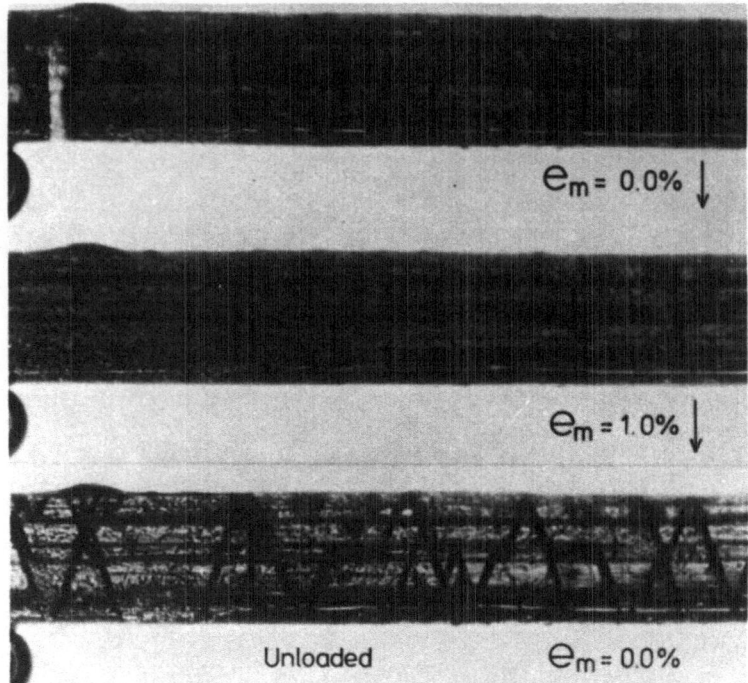

FIG. 17. Optical micrographs of twins in a polyDCHD single crystal fibre embedded in an epoxy resin cured at 100°C. The twins present in the undeformed state can be seen to disappear when the specimen is deformed to 1% strain and to reappear in different positions when the stress is removed. (Courtesy I. M. Robinson.)

chain direction therefore must first untwin the fibres before any significant stress is taken up by the fibres.[68] This is demonstrated in Fig. 17 where the twins are seen to disappear at about 1% strain and then re-form in different places when the specimen is unloaded.

The importance of resin shrinkage is further demonstrated in Fig. 18 where data are given for the axial strains, measured using Raman spectroscopy, at the mid-points of polyDCHD fibres in an epoxy resin matrix as a function of both tensile and compressive matrix strain. Examples are given for samples cured at both room temperature and 100°C. It can be seen that for tensile deformation the fibre and matrix strains are approximately equal for the room temperature cured sample which is not twinned, but in the case of the sample cured at 100°C the fibre

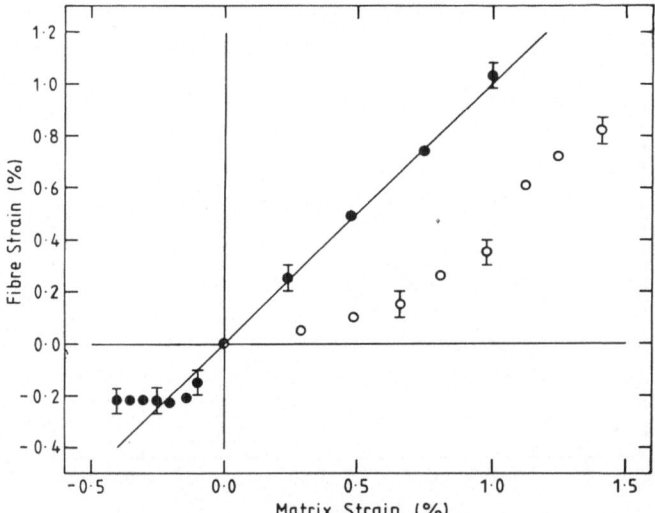

FIG. 18.   Axial strain at the midpoint of a polyDCHD fibre in an epoxy matrix as a function of tensile (positive) and compressive (negative) matrix strain. The closed circles represent the data for a specimen cured at room temperature.[68] The open circles represent data for a specimen cured additionally at 100°C.[83]

strain falls below that in the matrix due to the presence of twins.[68] It was also possible to determine the critical strain for twin formation in polyDCHD by loading the room-temperature cured sample in compression (Fig. 18). The fibre strain was found to be compressive as the Raman frequency was first observed to increase. The fibre strain approximately followed that of the matrix until at 0·2% compressive strain there was an abrupt levelling off with no further increase. At the same point twin bands were seen to appear on the fibre at this critical level of strain.[68]

The application of the Raman strain-measuring technique to these model polydiacetylene/epoxy composites is of general interest in the study of polymer composites. For example, measurement of the level of tensile strain that must be applied to remove the twins gives a direct *in situ* determination of the amount of shrinkage for a given resin system cured at a particular temperature. Also, since the twinning is related to the kinking that has been observed in other high-modulus fibres such as aromatic polyamides in composites,[87] the behaviour of polydiacetylene composites can be related directly to such systems of practical and commercial importance.

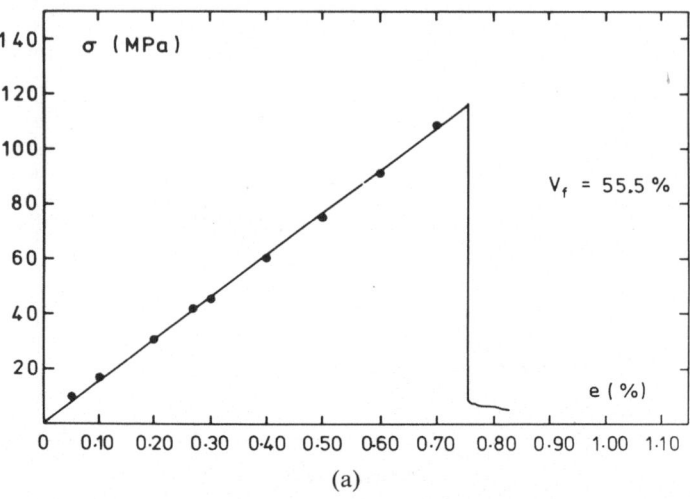

(a)

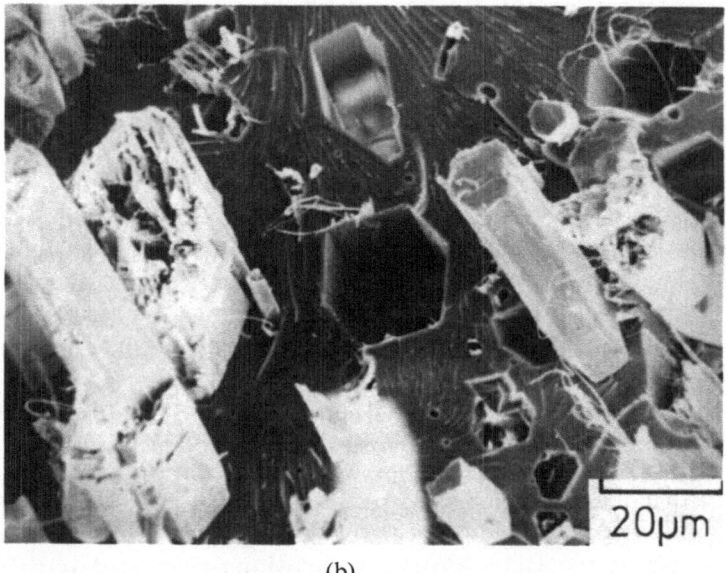

(b)

FIG. 19.   PolyDCHD/epoxy composites. (a) Stress/strain curve for a composite containing a volume fraction of 55·5% of fibres. (b) Scanning electron micrograph of a fracture surface of a composite specimen. (After Reference 84 by permission of the publishers, Chapman and Hall.)

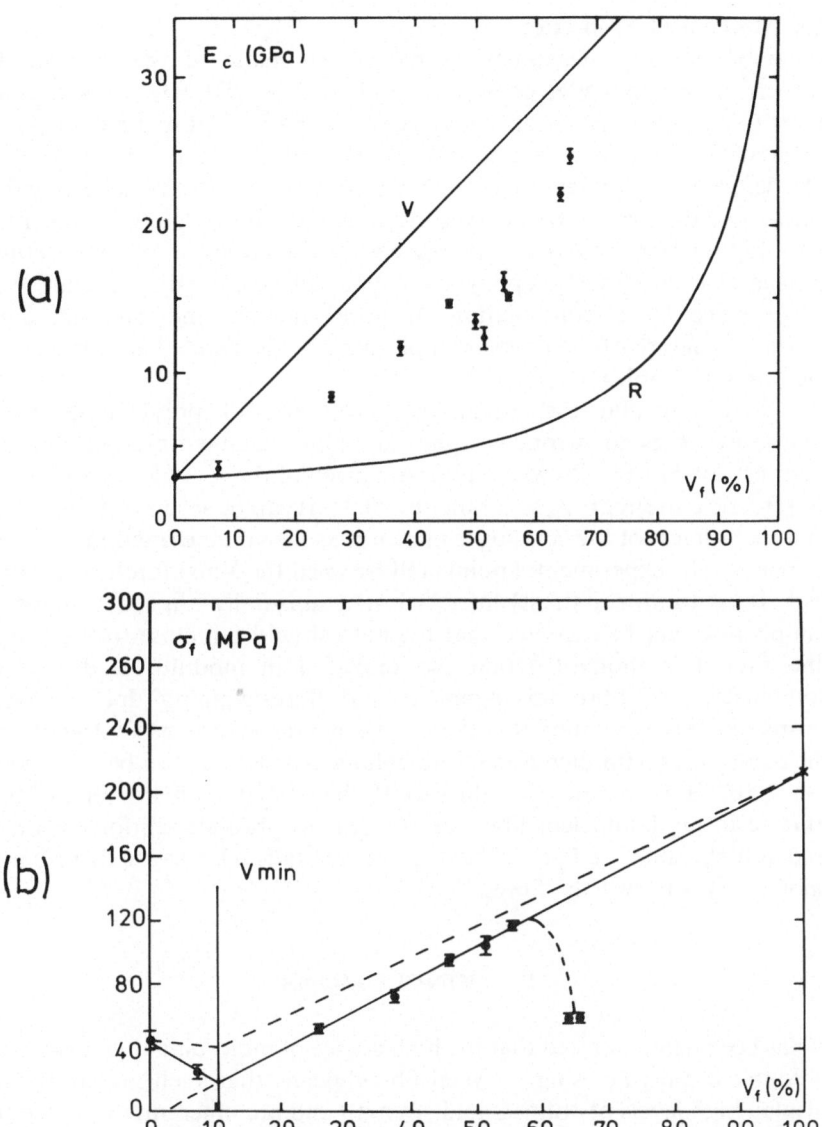

FIG. 20. Mechanical properties of the polyDCHD/epoxy composite system. (a) Variation of the Young's modulus of the composites with volume fraction of fibres; the lines for the Voigt (V) and Reuss (R) averages are shown. (b) Variation of the fracture strength of the composites with fibre volume fraction. (After Reference 84 by permission of the publishers, Chaopman and Hall.)

## 7.3. Mechanical Properties

Some preliminary measurements have been made of the mechanical properties of composites consisting of aligned polyDCHD single crystal fibres (~10 mm long) in an epoxy resin matrix.[84] Figure 19(a) shows a stress/strain curve for a sample with a fibre volume fraction, $V_f$, of 55·5%. The deformation is linear up to the fracture strain of ~0·8%, reflecting the linearity of the stress/strain curves for individual fibres at low strains (Fig. 9(a)). Figure 19(b) shows a scanning electron micrograph of the fracture surface of a polyDCHD/epoxy sample; it can be seen that fracture has taken place by a combination of fibre fracture and pull-out. The crystalline nature of the fibres is emphasized by the faceted appearance of the fibres and holes.

It is found that the mechanical properties of polyDCHD/epoxy composites depend strongly upon the fibre volume fraction; this is demonstrated in Fig. 20 where plots are given of the Young's modulus, $E$, and fracture strength, $\sigma_f$, as a function of $V_f$. It can be seen from Fig. 20(a) that the stiffness of the composite increases as the volume fraction of fibres increases. The experimental points fall between the Voigt (uniform strain) and Reuss (uniform stress) lines.[80] For a uniaxially aligned composite sample it would be expected that the data should fall close to the Voigt line, and it is thought[84] that the short-fall in modulus is due to a combination of fibre misalignment and fibre-twinning due to resin shrinkage. It is also found that there is a general increase in the strength of the composites with increasing fibre volume fraction, as can be seen from Fig. 20(b). However, at low values of $V_f$ the strength is below that of the pure resin until sufficient fibres are present to produce reinforcement,[76] and at high values of $V_f$ (>60%) the strength falls off because there is not enough resin to wet the fibres.[84]

# 8.  CONCLUSIONS

It has been demonstrated that the high degree of molecular alignment that is found in polymer single crystal fibres means that such materials can display high levels of stiffness and strength and are inherently resistant to creep. As a consequence of this it has been shown that composites produced by incorporating polydiacetylene single crystal fibres in an epoxy resin matrix have promising mechanical properties.

As well as enabling the development of a new type of reinforcing fibre, the study of these polymer single crystal fibres has led to significant advances in our understanding of the structure/property relationships in polymers. Examination of the structure of these materials using electron microscopy has enabled a detailed study of defects such as dislocations to be made, and it has also led to the development of high-resolution techniques allowing molecular detail to be seen at a level hitherto unobtainable with polymers. The characterization of surface defects such as steps has enabled estimates to be made of the theoretical strength of fibres and hence the strength of individual molecules. It has also been shown that polymer single crystal fibres undergo a newly discovered mode of twinning whereby the molecules bend across the twin boundary. The observation that the frequencies of the Raman active main-chain stretching modes are a strong function of externally applied strain has allowed the fibres to be used to follow in detail the micromechanics of fibre reinforcement in composites.

As well as leading to the development of a new class of reinforcing fibre, these studies on polydiacetylene single crystal fibres have helped to open up new areas of research in more conventional polymer systems. For example, there has recently been an upsurge of interest in the application of high-resolution transmission electron microscopy to a wide range of polymer crystals.[88] Also it has been found that the Raman strain measurement technique is applicable to other types of high modulus polymer fibres such as aromatic polyamides,[89] and this will allow detailed studies to be made of the micromechanics of reinforcement of composite systems reinforced with these fibres.

## ACKNOWLEDGEMENTS

Most of the work described in this chapter was carried out at Queen Mary College and the author is grateful to Professor D. Bloor and Dr D. N. Batchelder for introducing him to polydiacetylenes and for their continued encouragement and help. He would also like to thank Mr D. Ando and Mr I. F. Chalmers for their help with the preparation of monomers. Finally he must extend his gratitude to Dr R. T. Read, Dr C. Galiotis, Dr P. H. J. Yeung and Mr I. M. Robinson who performed the bulk of the experimental work described above.

## REFERENCES

1. ANDREWS, J. M. and WARD, I. M. (1970). *J. Mater. Sci.*, **5**, 411.
2. CAPACCIO, G. and WARD, I. M. (1975). *Polym. Eng. Sci.*, **15**, 219.

3. CAPACCIO, G., GIBSON, G. and WARD, I. M. (1979). In: *Ultra-high modulus polymers* (Ed. A. Ciferri and I. M. Ward), Applied Science Publishers, London, p. 1.
4. ZACHARIADES, A. E., MEAD, W. T. and PORTER, R. S. (1979). In: *Ultra-high modulus polymers* (Ed. A. Ciferri and I. M. Ward), Applied Science Publishers, London, p. 77.
5. MEAD, W. T., DESPER, C. R. and PORTER, R. S. (1979). *J. Polym. Sci. Polym. Phys. Ed.*, **17**, 859.
6. LEMSTRA, P. J. (1987). In: *Developments in Oriented Polymers—2*, chapter 2 (Ed. I. M. Ward), Elsevier Applied Science Publishers, London.
7. PENNINGS, A. J. and MEIHUIZEN, K. E. (1979). In: *Ultra-high modulus polymers* (Ed. A. Ciferri and I. M. Ward), Applied Science Publishers, London, p. 117.
8. SCHAEFGEN, J. R., BAIR, T. I., BALLOU, J. W., KWOLEK, S. L., MORGAN, P. W., PANAR, M. and ZIMMERMAN, J. (1979). In: *Ultra-high modulus polymers* (Ed. A. Ciferri and I. M. Ward), Applied Science Publishers, London, p. 173.
9. WOLFE, J. F., LOO, B. H. and ARNOLD, F. E. (1981). *Macromolecules*, **14**, 915.
10. ALLEN, S. R., FARRIS, R. J. and THOMAS, E. L. (1985). *J. Mater. Sci.*, **20**, 2727.
11. WEGNER, G. (1977). *Pure Appl. Chem.*, **49**, 443.
12. WEGNER, G. (1969). *Z. Naturforsch.*, **24b**, 84.
13. BLOOR, D., KOSKI, L., STEVENS, G. C., PRESTON, F. H. and ANDO, D. J. (1975). *J. Mater. Sci.*, **10**, 1678.
14. BLOOR, D. (1976). *Chem. Phys. Lett.*, **42**, 174.
15. LOCHNER, K., BÄSSLER, H., TIEKE, B. and WEGNER, G. (1978). *Phys. Stat. Sol.*, **B88**, 635.
16. DONOVAN, K. J. and WILSON, E. G. (1981). *Phil. Mag.*, **B44**, 9.
17. MOVAGHAR, B., MURRAY, D. W., DONOVAN, K. D. and WILSON, E. G. (1984). *J. Phys. C*, **17**, 1247.
18. BLOOR, D. (1985). *Phil. Trans. Roy. Soc.*, **A314**, 51.
19. ANDREWS, E. H. and MARTIN, G. E. (1973). *J. Mater. Sci.*, **8**, 1315.
20. PATELL, Y. R. and SCHULTZ, J. M. (1973). *J. Macromol. Sci. Phys.*, **B7**, 445.
21. NAKANISHI, H., SUZUKI, Y., SUZUKI, F. and HASGAWA, M. (1969). *J. Polym. Sci. A1*, **7**, 753.
22. KELLY, A. and GROVES, G. W. (1970). *Crystallography and crystal defects*, Longman, London.
23. DUDLEY, M., SHERWOOD, J. M., BLOOR, D. and ANDO, D. (1982). *J. Mater. Sci. Lett.*, **1**, 479.
24. SCHERMANN, W., WILLIAMS, J. O., THOMAS, J. M. and WEGNER, G. (1975). *J. Polym. Sci. Polym. Phys. Ed.*, **13**, 753.
25. YEE, K. C. and CHANCE, R. R. (1978). *J. Polym. Sci. Polym. Phys. Ed.*, **16**, 431.
26. GALIOTIS, C., READ, R. T., YEUNG, P. H. J., CHALMERS, I. F. and BLOOR, D. (1984). *J. Polym. Sci. Polym. Phys. Ed.*, **22**, 1589.
27. GALIOTIS, C., YOUNG, R. J., ANDO, D. J. and BLOOR, D. (1983). *Makromol. Chem.*, **184**, 1083.
28. GALIOTIS, C., YOUNG, R. J. and BATCHELDER, D. N. (1983). *J. Polym. Sci. Polym. Phys. Ed.*, **21**, 2483.
29. GALIOTIS, C. and YOUNG, R. J. (1983). *Polymer*, **24**, 1023.
30. STEVENS, G. C. and BLOOR, D. (1976). *Chem. Phys. Lett.*, **40**, 37.
31. BAUGHMAN, R. H. (1978). *J. Chem. Phys.*, **68**, 3110.

32. WENZ, G. and WEGNER, G. (1982). *Makromol. Chem., Rapid Commun.*, **3**, 231.
33. BAUGHMAN, R. H., GLEITER, H. and SENDFELD, N. (1975). *J. Polym. Sci. Polym. Phys. Ed.*, **13**, 1871.
34. READ, R. T. and YOUNG, R. J. (1979). *J. Mater. Sci.*, **14**, 1968.
35. KOBELT, D. and PAULUS, E. F. (1974). *Acta Cryst.*, **B30**, 232.
36. HADICKE, E., MEZ, H. C., KRAUCH, C. H., WEGNER, G. and KAISER, J. (1971). *Angew. Chem.*, **83**, 253.
37. APGAR, P. A. and YEE, K. C. (1978). *Acta Cryst.*, **B34**, 957.
38. WUNDERLICH, B. (1973). *Macromolecular physics*, Vol. 1, Academic Press, London.
39. GRUBB, D. T. (1974). *J. Mater. Sci.*, **9**, 1715.
40. READ, R. T. and YOUNG, R. J. (1984). *J. Mater. Sci.*, **19**, 327.
41. READ, R. T. and YOUNG, R. J. (1981). *J. Mater. Sci.*, **16**, 2922.
42. FRYE, C. J., WARD, I. M., DOBB, M. G. and JOHNSON, D. J. (1982). *J. Polym. Sci. Polym. Phys. Ed.*, **20**, 1677.
43. DOBB, M. G., JOHNSON, D. J. and SAVILLE, B. P. (1980). *Phil. Trans. Roy. Soc.*, **A294**, 483.
44. YOUNG, R. J. and YEUNG, P. H. J. (1985). *J. Mater. Sci. Lett.*, **4**, 1327.
45. YEUNG, P. H. J. and YOUNG, R. J. (1986). *Polymer*, **27**, 202.
46. KLUG, A. and DEROSIER, D. J. (1966). *Nature*, **212**, 29.
47. LINDENMEYER, P. H. (1966). *J. Polym. Sci.*, **C15**, 109.
48. HOLLAND, V. F. and LINDENMEYER, P. H. (1965). *J. Appl. Phys.*, **36**, 3049.
49. PETERMANN, J. and GLEITER, H. (1972). *Phil. Mag.*, **25**, 813.
50. YOUNG, R. J. and PETERMANN, J. (1982). *J. Polym. Sci. Polym. Phys. Ed.*, **20**, 961.
51. YOUNG, R. J., READ, R. T. and PETERMANN, J. (1981). *Inst. Phys. Conf. Ser.*, No. 61, Ch. 10, p. 475.
52. YOUNG, R. J., READ, R. T. and PETERMANN, J. (1981). *J. Mater. Sci.*, **16**, 1835.
53. BATCHELDER, D. N. and BLOOR, D. (1979). *J. Polym. Sci. Polym. Phys. Ed.*, **17**, 569.
54. GALIOTIS, C. (1982). Ph.D. Thesis, University of London.
55. KINLOCH, A. J. and YOUNG, R. J. (1983). *Fracture behaviour of polymers*, Applied Science Publishers, London.
56. FRANK, F. C. (1970). *Proc. Roy. Soc.*, **A319**, 127.
57. YOUNG, R. J. (1981). *Introduction to polymers*, Chapman and Hall, London.
58. ODAJIMA, A. and MAEDA, T. (1966). *J. Polym. Sci.*, **C15**, 55.
59. TRELOAR, L. R. G. (1960). *Polymer*, **1**, 95.
60. LEYRER, R. J., WEGNER, G. and WETTLING, W. (1978). *Ber. Bunsenges, Phys. Chem.*, **82**, 697.
61. ENKELMANN, V., LEYRER, R. J., SCHLEIER, G. and WEGNER, G. (1980). *J. Mater. Sci.*, **15**, 168.
62. REHWALD, W., VONLANTHEN, A. and MEYER, W. (1983). *Phys. Stat. Sol.*, (a) **75**, 219.
63. BOWDEN, P. B. and YOUNG, R. J. (1974). *J. Mater. Sci.*, **9**, 2034.
64. YOUNG, R. J., BLOOR, D., BATCHELDER, D. N. and HUBBLE, C. L. (1978). *J. Mater. Sci.*, **13**, 62.
65. YOUNG, R. J., DULNIAK, R., BATCHELDER, D. N. and BLOOR, D. (1979). *J. Polym. Sci. Polym. Phys. Ed.*, **17**, 1325.

66. WILDING, M. A. and WARD, I. M. (1978). *Polymer*, **19**, 969.
67. WOODS, D. W., BUSFIELD, W. K. and WARD, I. M. (1985). *Plast. Rubb. Process. Appln.*, **5**, 157.
68. ROBINSON, I. M., YEUNG, P. H. J., GALIOTIS, C., YOUNG, R. J. and BATCHELDER, D. N. (1986). *J. Mater. Sci.*, **21**, 3440.
69. DOBB, M. G., JOHNSON, D. J. and SAVILLE, B. P. (1981). *Polymer*, **22**, 960.
70. KELLER, A., personal communication.
71. PIETRALLA, M. (1976). *Colloid Polym. Sci.*, **254**, 249.
72. BEVIS, M. (1978). *Colloid Polym. Sci.*, **256**, 234.
73. YOUNG, R. J. (1979). In: *Developments in polymer fracture* (Ed. E. H. Andrews), Applied Science Publishers, London.
74. BATCHELDER, D. N., personal communication.
75. BRENNER, S. S. (1962). *J. Appl. Phys.*, **33**, 33.
76. KELLY, A. (1966). *Strong solids*, Clarendon Press, Oxford.
77. MARSH, D. M. (1963). In: *Fracture in solids* (Ed. D. C. Drucker and J. J. Gilman), Interscience, New York.
78. KINLOCH, A. J. and YOUNG, R. J. (1983). *Fracture behaviour of polymers*, Applied Science Publishers, London.
79. KAUSCH, H. H. (1978). *Polymer Fracture*, Springer-Verlag, Berlin.
80. HULL, D. (1981). *An introduction to composite materials*, Cambridge University Press.
81. LADIZESKY, N. H. and WARD, I. M. (1985). *Pure Appl. Chem.*, **57**, 1641.
82. GREENWOOD, J. H. and ROSE, P. G. (1974). *J. Mater. Sci.*, **9**, 1809.
83. GALIOTIS, C., YEUNG, P. H. J., YOUNG, R. J. and BATCHELDER, D. N. (1984). *J. Mater. Sci.*, **19**, 3640.
84. YOUNG, R. J., GALIOTIS, C., ROBINSON, I. M. and BATCHELDER, D. N. *J. Mater. Sci.* (Submitted for publication.)
85. COX, H. L. (1952). *Brit. J. Appl. Phys.*, **3**, 72.
86. BATCHELDER, D. N. (1976). *J. Polym. Sci. Polym. Phys. Ed.*, **14**, 1235.
87. DETERESA, S. J., ALLEN, S. R., FARRIS, R. J. and PORTER, R. S. (1984). *J. Mater. Sci.*, **19**, 57.
88. THOMAS, E. L. (1985). *Polymer Prepr.*, **26**, 314.
89. GALIOTIS, C., ROBINSON, I. M., YOUNG, R. J., SMITH, B. E. J. and BATCHELDER, D. N. (1985). *Polymer Commun.*, **26**, 354.

*Chapter 2*

# HIGH-STRENGTH/HIGH-MODULUS STRUCTURES BASED ON FLEXIBLE MACROMOLECULES: GEL-SPINNING AND RELATED PROCESSES

P. J. Lemstra*, R. Kirschbaum

*DSM, Central Research, Geleen, The Netherlands*

T. Ohta and H. Yasuda

*Toyobo, Katata Research Institute, Ohtsu, Shiga, Japan*

## 1. INTRODUCTION

Amongst the various developments in the area of high-performance fibres, two major routes can be discerned which are completely different in respect of the starting (base) materials, respectively intrinsically rigid as opposed to intrinsically flexible macromolecules.

### 1.1. Rigid Chains

More than 50 years ago Carothers and Hill[1] formulated the conditions requisite for the production of a 'useful fibre', such as the necessity for long chain molecules which are perfectly ordered in an array with the chain axis parallel to the fibre direction. Along these lines the design and development of intrinsically stiff macromolecules forms, at least in retrospect, a logical approach to the exploitation of the intrinsic possibilities with respect to stiffness and strength of a covalently bonded

* *Present address:* Department of Polymer Technology, Eindhoven University of Technology, Eindhoven, The Netherlands.

array of atoms, since chain-extension and alignment are achieved comparatively easily (not taking into account the sophisticated chemical work done beforehand) in comparison with flexible macromolecules.

Examples of really rigid chain molecules are PBT (poly-$p$-phenylene benzobisthiazole) and its molecular analogue PBO (poly-$p$-phenylene benzobisoxazole). In both cases the persistence length, $P$, measured by light scattering in dilute solutions is, within experimental error, identical with the contour length, $L$, indicative of a 'rigid rod nature'.[2] Typical tensile properties[3] of heat-treated PBT fibres spun from solution are: tenacity 3·5 GPa, modulus 250GPa and elongation at break 1·2%, or the specific values for strength and stiffness, viz. 2·2 N tex$^{-1}$ (see Table 4 for conversion of units) and 160 N tex$^{-1}$ respectively (density of PBT is $\sim 1\cdot 6\,\mathrm{g\,cm}^{-3}$).

The prime examples of rigid chain polymers however are the aromatic polyamides (aramids), notably poly($p$-phenylene terephthalamide) (PPTA), marketed under the trade names of Kevlar (DuPont) and Twaron (Akzo/Enka). The tensile properties of e.g. the Kevlar 49 grade are: tenacity 3 GPa, modulus 130 GPa and strain at break 2% or, in specific values, tenacity 2 N tex$^{-1}$ and modulus 90 N tex$^{-1}$ (the density of PPTA is $1\cdot 45\,\mathrm{g\,cm}^{-3}$).

Although the aromatic polyamides are not strictly rigid chains[3,4] (for PPTA the ratio $L/P$ is about 4), experimental conditions have been found to promote chain-extension and orientation in the fibre direction. These experimental conditions encompass spinning from liquid crystalline (nematic) solutions and optimized spinning and coagulation procedures followed by heat treatment under tension to promote and retain a high degree of chain-orientation/extension in the fibre direction to achieve optimum mechanical properties.[5]

## 1.2. Flexible Chains

Flexible (regular) chains tend to fold during crystallization (see Section 1.3), and consequently in conventional spinning methods folded-chain type crystals will be formed during solidification (melt-spinning), solvent evaporation (dry-spinning) or coagulation (wet-spinning). Of course this statement is an oversimplification of reality since, during spinning of stereo-irregular polymers such as poly(acrylonitrile), L–L phase separation takes place during coagulation and chain-folding is not very pronounced. Nevertheless, the tendency to form folded-chain crystals shown by crystallizable polymers such as the polyamides, polyesters, polypropylene and polyvinylalcohol during solidification from the melt or

TABLE 1

ELASTIC MODULI:[6,7] COMPARISON BETWEEN THEORETICAL AND ACTUAL VALUES

|  | $E_{actual}$ (GPa) | $E_{theor}$ (GPa) |
|---|---|---|
| Polyvinylalcohol | 30 | 200–250 |
| Polyester (PETP) | 20 | 120–150 |
| Polyamide (nylon-6) | 6 | 170–270 |
| Polypropylene | 12 | 35–49 |

solution is hindering for achieving chain-extension and formation of high-tenacity fibres by conventional spinning methods.

Transformation of the as-spun structures into chain-extended type crystals via well known post-drawing techniques, usually in a temperature range close to but below the melting point, is only partly successful, at least if we compare the values of axial (Young's) moduli of large-scale manufactured technical yarns with their theoretical limits (see Table 1).

The theoretical moduli and the actual values for technical yarns (Table 1) indicate a major gap between practical and theoretical limits. Comparison with the moduli of high-performance fibres based on rigid chains makes it clear that the conventional spinning techniques involving post-tensile drawing are highly inadequate for the production of high-strength/high-modulus structures based on flexible macromolecules. In Table 1 no reference is made to theoretical values for tenacity since this property, in contrast with low-strain properties such as moduli, is difficult to estimate for real (finite chain) systems (see below).

## 1.3. Polyethylene vs. Other Polymers

Soon after the discovery of linear polyethylene (Ziegler, 1954), an important feature was found: the phenomenon of folded-chain crystallization. Storks[8] had mentioned the possibility of chain folding during crystallization, on the basis of electron diffraction studies of thin films of gutta-percha in 1938. The concept of chain folding, however, remained unnoticed until about 1955, when polyethylene single crystals were obtained from dilute solutions.[9] Keller concluded from the electron diffraction pattern of PE single crystals that the direction of the chains was perpendicular to the basal plane, as shown schematically in Fig. 1.

Although the concept of chain-folded crystallization of PE and other regular/flexible macromolecules has fascinated many scientists up to now, scattered research activities had already started in the 1960s to pursue chain extension. Polyethylene featured as the prime candidate in these

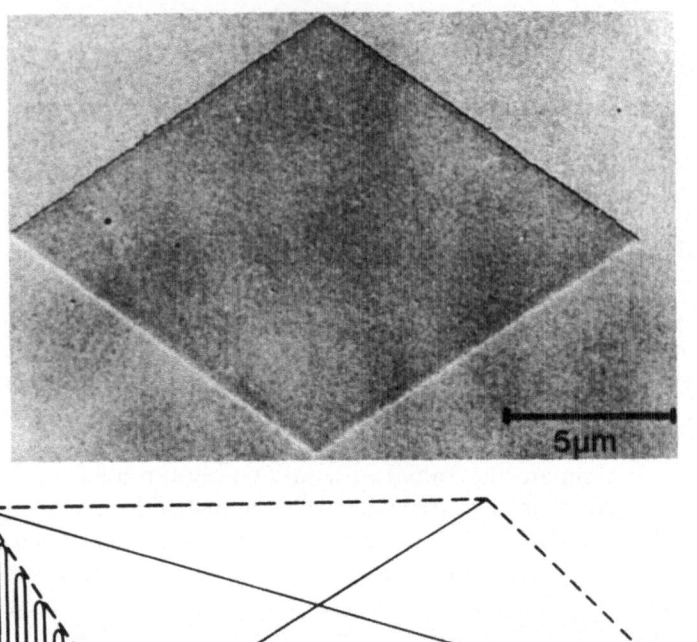

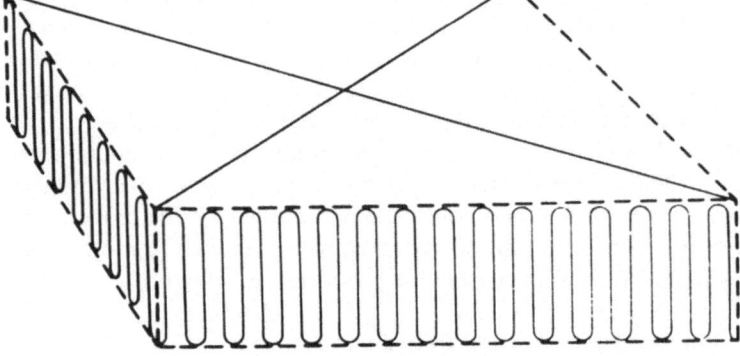

FIG. 1.    Electron micrograph of PE single crystal and schematic representation of
macromolecules within the crystal. (Courtesy of Prof. A. Keller, Bristol.)

studies, partly due to its simple chemical structure and its availability in a
wide range of chain lengths, but also due to its intrinsic possibilities.
Polyethylene can be synthesized with molecular weights exceeding
$10^6$ kg kmol$^{-1}$ and with a nearly linear structure. The zig-zag chain
conformation in the solid state and the absence of pendant groups implies
that the PE chain has a small cross-sectional area and consequently that
the number of load-bearing elements per fibre cross-section is high,
provided the chains are fully extended in the fibre direction. Theoretical
calculations showed that the strength of linear PE could be as high as
19–25 GPa,[10,11] with a corresponding modulus of about 300 GPa. One

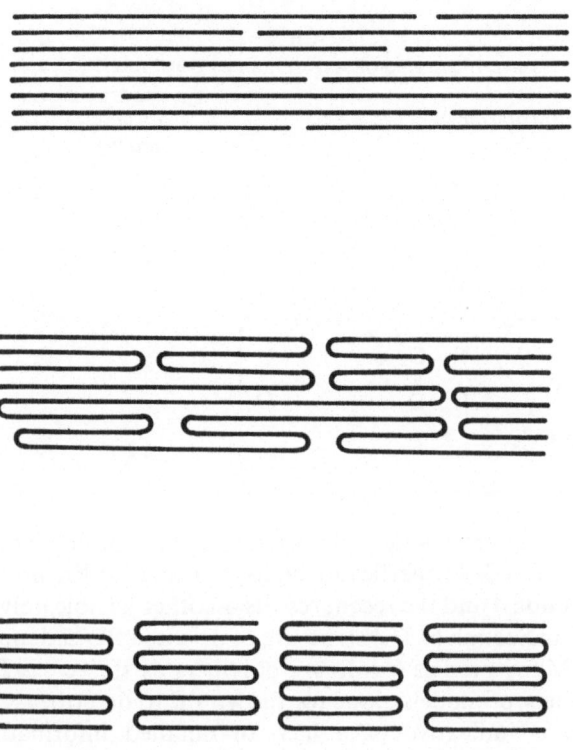

FIG. 2.   Chain-orientation vs. chain-extension.

has to be cautious in using these strength values because they were calculated from loading individual and infinite chains to the limit (rupture of C–C bonds). Actually, we have to deal with finite chains and the maximum tenacity values are determined by other factors such as chain length, degree of chain orientation, crystallinity and crystal size and, above all, chain extension and topology. This is shown schematically in Fig. 2 for three examples, all possessing nearly perfect chain orientation but having a very different degree of chain extension. It is clear without further comment that the tenacity increases from virtually zero to its maximum value.

### 1.4. Aim (Scope) of Review
The origin and development of various techniques and processes to produce high-strength/high-modulus PE structures will be reviewed, with emphasis on solution-spinning techniques in Section 2. The basic aspects

TABLE 2
SOME HIGH-PERFORMANCE FIBRES COMPARED

| Material | Specific tenacity ($N\ tex^{-1}$) | Specific modulus ($N\ tex^{-1}$) |
|---|---|---|
| PBT | 2·2 | 160 |
| PPTA | 2 | 90 |
| HP-PE $-\!\left[C\!-\!C\right]_n$ | 3·5 | 130 |
| PE (isotropic) | <0·05 | <2 |

and proposed mechanisms for ultra-drawing of polyethylene will be discussed in Section 3. Properties of high-performance PE fibres will be presented in Section 4 and the recent results on other flexible polymers will be summarized in Section 5. The structure of oriented PE and modelling is not discussed. Many models for drawn polymers exist (see e.g. Reference 12), and the variety of models alone for drawn PE to describe its structure and properties is indicative of a lack of detailed information on a molecular level. Standard techniques such as X-ray scattering, birefringence etc. reveal chain orientation but fail to provide adequate information about chain extension.

The techniques and processes discussed certainly do not constitute a recipe for making high-strength/high-modulus structures from flexible polymers. On the contrary, polyethylene is rather an exception amongst its class of flexible chain molecules and currently pre-marketed high-performance PE fibres (HP-PE) are exceptional if we compare their specific tensile properties with those of other high-performance fibres, the more so if we take into account its simple chemical structure in relation to properties (see Table 2).

## 2.   HISTORICAL SURVEY

Numerous publications exist on the drawing behaviour of polyethylene and the related crystal structure and morphology of oriented PE structures. It is impossible within the scope and purpose of this review to discuss and comment upon the various contributions in this field. Therefore a selection has been made of views, experimental results and techniques which are directly related to current processes for the production of high-strength/high-modulus PE fibres. It is, of course, in retrospect always possible to select those experimental results from the literature which, after being presented in chronological order, will give the impression that the development towards the production of high-strength/high-modulus PE fibres has been straightforward. This is however far from reality; in fact the route towards current processes has been a tortuous one and, as usual, serendipity has played an important role, as can be inferred from the literature survey presented below.

## 2.1. Early Attempts (before 1970)

### 2.1.1. Solution Routes

Solution-spinning of linear polyethylene dates back to 1956, when a patent application was filed by the Vereinigte Glanzstoff-Fabriken (now Akzo-Enka) concerning wet-spinning of polyolefins.[13] The inventor, Jurgenleit, reported spinning of linear polyethylenes from moderately concentrated solutions, 10–18%; after post-drawing the spun filaments, tenacities up to 1·2 GPa were obtained. Similar values for tenacity were found by Sato and Hirai in spinning concentrated PE solutions and subsequent drawing.[14]

Solution-spinning of ultra-high molecular weight (UHMW) PE ($M_w >$ $10^6$ kg kmol$^{-1}$) was performed by Zwick[15] from naphthalene solutions. No post-drawing was mentioned in his patent application, nor were fibre properties discussed. Blades and White[16] introduced their so-called flash-spinning technique. A pressurized solution of linear polyethylene (melt-index 0·5 and solution concentration 13% in halogenated hydrocarbons) was extruded at high speeds and temperatures (about 200°C). The fibrillated strand yarn which precipitated upon rapid cooling and almost instantaneous solvent evaporation was subjected to slow drawing. Maximum values for tenacity and modulus were reported to be about 1·4 GPa and 20 GPa, respectively.

### 2.1.2. Chain Extension in Solution

Mitsuhashi[17] was probably the first to attempt inducing chain extension directly in solution. Figure 3 shows his original drawing of a Couette type apparatus used in the early 1960s. Mitsuhashi reported the formation of fibrous 'string-like' PE structures upon stirring of linear polyethylene

(Marlex 50) in xylene. No mechanical properties of the fibrous crystal mats precipitated on the rotor surface were reported. His original observations remained practically unnoticed until about 10 years later a similar experimental set-up was used by Pennings et al. (see below) and the 'string-like' PE structures were baptized 'shish-kebab' structures.

Stirring polymer solutions to induce chain extension is less obvious than might be thought at first sight. In general, to stretch the chain the hydrodynamic driving force should exceed the entropic restoring force towards the random coil conformation. The type of flow is important and elongational flows are particularly suitable in view of the aim of fully stretching out an individual macromolecule.[18] The following conditions are claimed to hold for chain extension:[19]

$$\dot{\varepsilon}.\tau > 1; \qquad \dot{\varepsilon}.\Delta t > 1 \tag{1}$$

In eqn (1), $\dot{\varepsilon}$ is the strain rate, $\tau$ the characteristic relaxation time and $\Delta t$ the time during which the macromolecule is exposed to the flow field. The relaxation time is strongly dependent on molecular weight.

Mackley and Keller[20] estimated for polyethylene in dilute solutions that $\tau \approx M^{1.75}$, and recent studies[21] for monodisperse polystyrene gave $\tau \approx M^{1.5}$. In dilute solutions at polymer concentrations $\phi \geq \phi^*$ ($\phi^*$ is the overlap concentration), and in good solvent systems, only fractional chain extension will be obtained in the case of linear polyethylene due to the broad molecular weight distribution. It was calculated[20] for the system polyethylene in xylene that at a strain rate of $1000\,\mathrm{s}^{-1}$ only molecules with molar masses $> 2 \times 10^6$ will become fully stretched out. The extremely high strain rates require special devices for chain extension.[21]

The observation of Mitsuhashi that in a simple Couette type apparatus oriented PE structures could be produced is, in retrospect, due to two main factors. In semi-dilute solutions (in the case of Mitsuhashi $\phi > 2\%$) entanglement-coupling plays an important role and induced chain extension can be made permanent if the solution is slightly supercooled by nucleation and alignment of (partly) oriented chains (to be discussed below).

### 2.1.3. Single Crystal Drawing
Fundamental drawing studies, involving PE single crystals, were undertaken by Statton[22] and Maeda et al.[23,24] It was reported that single crystal mats of PE were remarkably ductile at elevated temperatures, 100–120°C, i.e. close to but below the melting point. Later studies by Barham and Keller[25] showed that PE single crystal mats could be drawn up to 40 × with corresponding moduli of about 50 GPa. The main

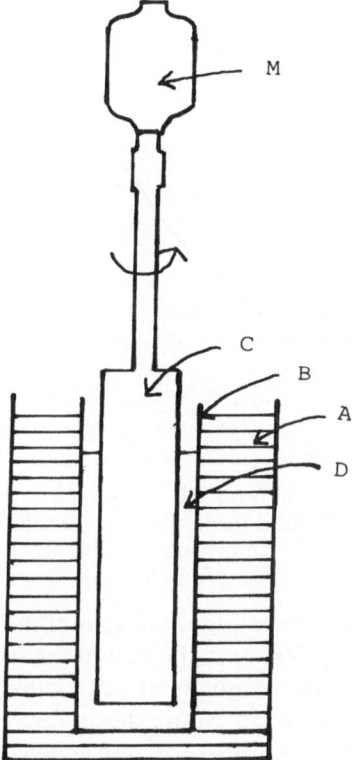

FIG. 3. Stirring-induced crystallization: M, motor; A, oil thermostat; B, cylinder containing PE solution (D); C, rotor. (From original photograph by S. Mitsuhashi.)

conclusion from Statton's original work is that folded-chain crystals grown from dilute solutions are ductile at elevated temperatures and unfold rather easily in the direction of the applied stress, an observation which was comprehended more than ten years later to its full extent after the invention of the gel-spinning process.

## 2.2. Developments during 1970–80
### 2.2.1. Deformation in the Solid State
In 1970 Andrews and Ward[26] started systematic drawing studies of spun/melt crystallized PE filaments. They reported that the axial Young's modulus of the drawn filaments increased with increasing draw ratio and moduli of up to about 20 GPa. Later, Ward et al.[27] studied the influence of

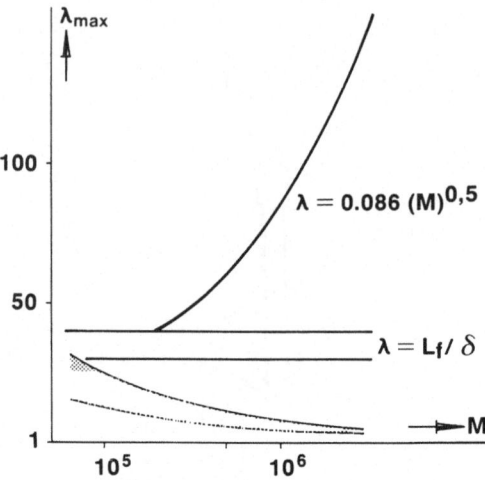

FIG. 4. Maximum extensibility of PE macromolecules and maximum observed draw ratio of melt-crystallized PE in the solid state (shaded area) as a function of molecular weight, $M$.

spinning and quenching/cooling conditions of the as-spun filaments on the maximum drawability. Via optimization of the spinning and solidification conditions, fibres could be obtained possessing moduli of up to about 70 GPa, or specific values exceeding the stiffness of glass and steel. Melt-spinning followed by ultra-drawing in a temperature range close to but below the melting temperature is however limited to certain molecular ranges, for two reasons:

(a)  With increasing molecular weight, melt-spinning becomes increasingly difficult due to a strong rise in melt viscosity, and related phenomena such as melt fracture are prohibitive to continuous in-line spinning/drawing. An upper limit is an $M_w$ of about $5 \times 10^5$.

(b)  With increasing molecular weight, melt-crystallized polyethylene increasingly resists deformation in the (semi-)solid state, i.e. post-drawing of melt-spun filaments becomes less adequate. The decrease in drawability/draw ratio with increasing molecular weight was studied systematically by Ward and coworkers and Fig. 4 shows their experimental results for the maximum draw ratio as a function of molecular weight of melt-crystallized polyethylenes (shaded area in Fig. 4). A limiting value for $\lambda_{max}$ is obtained below 10, for $M_w \geq 10^6$ kg kmol$^{-1}$, sometimes referred to as a natural draw ratio.

The concept of a natural draw ratio has been established in tensile drawing of melt-spun filaments, for instance polyamides with $\lambda_{max} \sim 5$–6. However, a limiting natural draw ratio is not predicted by any current model of polymer crystalline structure/morphology. The maximum extension ratio of chain molecules, which equals the extended chain length, $L$, divided by the end-to-end distance, can be estimated provided that the chain conformation is known. In the case of melt-crystallized polymers, the topology of chains is still very much a matter of debate. Although detailed studies on the drawing behaviour of polyethylene as a function of molecular weight, crystallization conditions etc. could provide interesting additional information with respect to the organization of chains in the solid state, we will simplify the actual complex situation by stating that in the case of melt-crystallized PE the chain conformation should lie between the two extremes (a) regularly folded (adjacent re-entry) and (b) random coils (Erstarrungsmodell[28]). In the case of regularly folded chain crystals the maximum molecular extension ratio is given by the ratio of the fold length, $L_f$, and the chain diameter, $\delta$:

$$\lambda = L_f/\delta \tag{2}$$

and for random coils:[29]

$$\lambda = 0.086(M)^{0.5} \tag{3}$$

As can be inferred from Fig. 4, the maximum molecular extension ratios are much higher than the maximum observed macroscopic draw ratios (shaded area). In the drawing of solid (flexible) polymers the macroscopic deformation correlates with the molecular extension,[30] and it is clear from the calculated extension ratios that the decrease in drawability with $M$ for polyethylene is not related to the individual chains but to other factors, which will be discussed in Section 3.

Despite its limitations with respect to molecular weight, reflected in the ultimate strength values of about 1 GPa, melt-spinning/drawing of linear PE has advantages, such as its simplicity and consequently low cost price of the fibres and tapes, and, in terms of price performance, is attractive for many applications.[31]

Deformation in the solid state is not restricted to tensile drawing of spun filaments. Solid-state extrusion (hydrostatic extrusion) of pre-formed PE billets has been performed.[32,33] In such processes one avoids the melt-extrusion/spinning step but, as is clear from the discussion above, in deforming solid PE the same limitations are encountered with respect to molecular weight. In order to obtain high-modulus structures, high EDR's (extrusion draw ratios) are necessary, comparable to draw ratios in tensile drawing. More recently, the method of drawing through suitably profiled dies (die-drawing) has been reported as a means of producing three-dimensional oriented structures such as tubes and sheets.[34]

### 2.2.2. Solution Routes

In the mid-1960s it was found at DSM Central Research that fibrous PE crystals are formed in a Couette apparatus if the inner rotor exceeds a

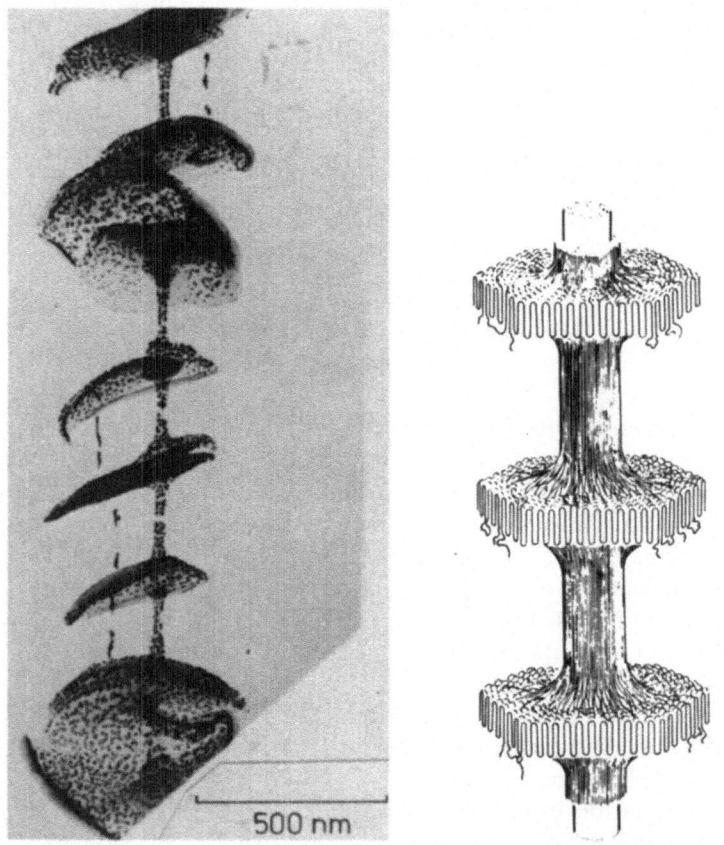

500 nm

FIG. 5.    Electron micrograph of PE 'shish-kebab' and schematic representation of macromolecular organization. (Courtesy of Dr M. Hill, Bristol.)

critical rotation rate coinciding with the onset of Taylor vortices.[35] The correlation between the onset of Taylor vortices, which are known to contain elongational components, and the formation of fibrous crystals became less pronounced when high molecular weight PE was employed. As mentioned before in discussing Mitsuhashi's work, formation of fibrous crystal structures in stirring semi-dilute, high molecular weight PE solutions is in part due to entanglement coupling, as became clear later after detailed studies on surface growth and gel-spinning (see below).

The morphology of stirring-induced fibres has become known as the 'shish-kebab', a central core consisting of more or less extended PE molecules (shish) and folded-chain type crystals deposited on the core (kebabs); see Fig. 5. For a detailed discussion of the formation of shish-kebabs the reader is referred elsewhere.[19,36,37]

The structure of shish-kebab type fibrous polyethylene is far from the ideal arrangement of PE macromolecules with respect to stiffness and strength, and reflects once more the problem of obtaining full chain-extension. Due to the presence of lamellar overgrowth, moduli of precipitated fibrous PE 'shish-kebabs' were limited to up to about 25 GPa,[38] to be compared with >50 GPa in the cases of melt-spinning/drawing and single-crystal drawing.

The work on stirring-induced crystallization was followed by various techniques such as free growth,[39] culminating in the so-called surface-growth technique.[40] Figure 6 shows a schematic representation of this technique. A seed fibre is immersed in a dilute solution of UHMW-PE and from the surface of the rotating inner-cylinder fibrous, tape-like polyethylene structures can be withdrawn. Under optimized experimental

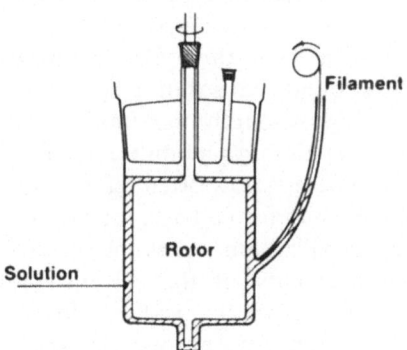

FIG. 6.   Schematic representation of surface-growth technique. (Taken from Reference 40.)

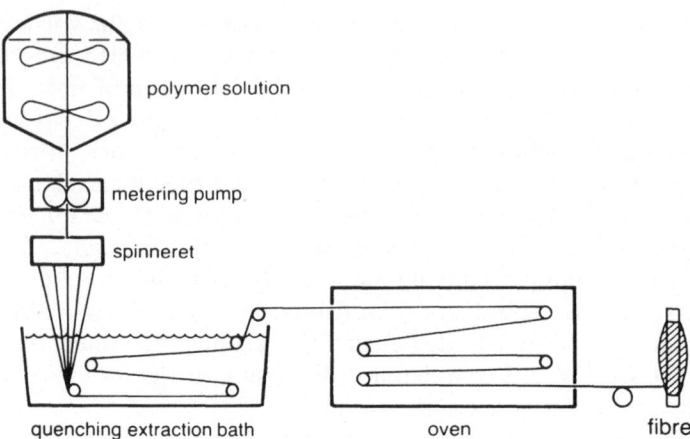

polymer solution

metering pump

spinneret

quenching extraction bath          oven          fibre

FIG. 7.   Schematic representation of gel-spinning process.

conditions with respect to solution concentration, temperature and take-up speed, oriented PE structures could be obtained possessing tensile moduli of over 100 GPa and strength values above 3 GPa. This technique was studied in detail by Zwijnenburg, Meihuizen, and Pennings (University of Groningen)[41] and by Barham and Keller (University of Bristol),[37] and in fact represents a milestone since it provided the experimental proof that high-modulus/high-strength PE structures based on UHMW-PE were possible. Although a lot of effort was devoted to understand the paramount factors governing the speed and production of oriented PE in the surface-growth technique, a technological process was not feasible in view of the limited take-up speeds, which were lower than $1 \, m \, min^{-1}$.

At the end of the 1970s a remarkable observation was made at DSM Central Research. It was found that solution-spun UHMW-PE filaments could be ultra-drawn to high-strength/high-modulus fibres, with tenacity $>3$ GPa and corresponding Young's moduli of over 100 GPa.[42] Figure 7 shows schematically this process, now often referred to as gel-spinning. A solution of UHMW-PE is spun into a bath, for instance water, and upon cooling solidification/crystallization takes place producing a gel-like appearance due to immiscibility of the solvent with water. The as-spun/quenched filaments are mechanically sufficiently stable to be transported via a roller system into an oven in which super-drawing is performed at elevated temperatures. At first glance, gel-spinning seems identical with standard solution-spinning. The remarkable feature,

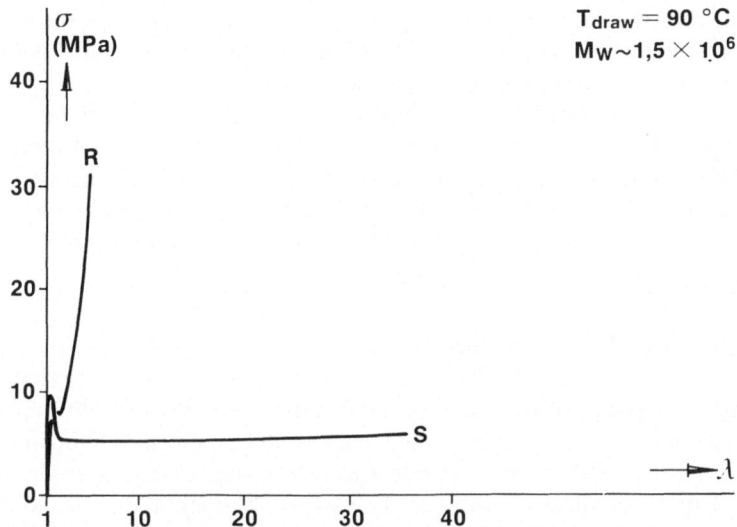

FIG. 8. Stress–strain behaviour (nominal stress) of respectively melt-crystallized (R) and solution-crystallized (S) UHMW-PE; the solution-cast film (initial polymer conc. 2%) was dried and extracted.

however, is that even after complete solvent removal from the as-spun filament, for instance by extraction, ultra-drawing is still possible. Figure 8 demonstrates the difference in drawability between melt-crystallized and solution-spun/cast UHMW-PE in the solid state, and Fig. 9 shows the effectiveness of the tensile drawing procedure. Figure 9 presents the data from the original patent applications[42] based on isothermal drawing, but

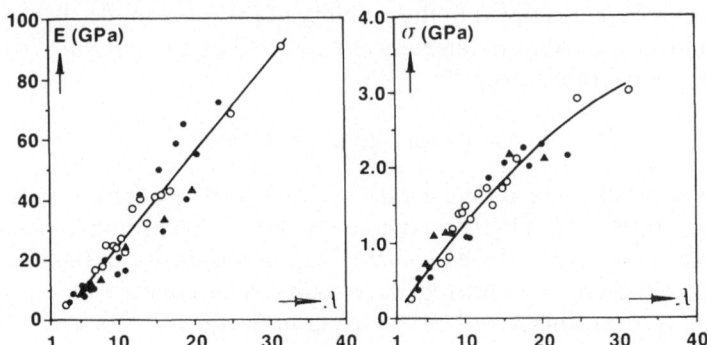

FIG. 9. Effectiveness of drawing gel-filaments: ○, 'wet' filaments (90% residual solvent); ●, partly dried filaments (6% residual solvent); ▲, dried (extracted) filaments; drawing performed single-stage at 120°C.

through optimization of the drawing procedure, such as multi-stage drawing, tenacities of between 3 and 4 GPa and moduli of up to 150 GPa can be realized by in-line spinning/drawing; see Section 4.

The striking observation is that HMW-PE, $M_w > 5 \times 10^5$, can be ultra-drawn after spinning from solution, even after *complete* solvent removal from the as-spun filaments. In this respect there is a sharp contrast with melt-crystallized HMW-PE's (see Fig. 4). The major conclusion is that the 'constraints' which limit the drawability in melt-crystallized PE are removed by spinning/casting from solution.

### 2.3. Recent Developments (since 1980)

Since the invention of the gel-spinning process, several (patent) applications have appeared describing variations of this process, but the basic principles have not been challenged, as will be discussed in Section 3.

Spinning of UHMW-PE solutions into extracting solvents was claimed to be essential for ultra-drawing of the as-spun filaments, due to induced porosity promoting drawability. In a recent review on advances in high-strength fibres[43] it is stated that non-porous fibres in fact cannot be drawn to high draw ratios. In their patent applications, Kavesh and Prevorsek[44] describe the spinning of UHMW-PE solutions and extraction of the solvent from the as-spun filaments with, in their terminology, a volatile solvent, leading to formation of a porous so-called 'xero-gel' fibre. These 'xero-gel' fibres are drawn in a two-stage process and mechanical properties are attained which are similar in magnitude to those obtained in 'standard' gel-spinning. The concept of porosity influencing the drawability was also discussed by Kalb and Pennings.[45] The drawability of as-spun fibres is related to the mobility of the macromolecules, which is determined by various parameters as expressed by the simplified equation for non-recoverable creep:[46]

$$\dot{\varepsilon} = A \exp[-(\Delta E - \sigma \Delta V)/kT] \tag{4}$$

In eqn (4), $\dot{\varepsilon}$ is the creep rate, $\Delta E$ the activation energy for transport, $\sigma$ the applied stress and $\Delta V$ the activation volume. According to Kalb and Pennings an increase in free volume, the microporosity in the fibre to be drawn, will lead to improved drawability. The concept of porosity in relation to drawability will be discussed in Section 3.

A seemingly new method was introduced by Mitsui, the so-called melt-kneading process.[47] A mixture of UHMW-PE and paraffin wax (melting point between 40 and 120°C) was extruded/spun, and after quenching in water the as-spun/quenched filaments or tapes were drawn in, for instance, n-decane at 130°C. Tenacities of the drawn fibres and tapes of up to 2·4 GPa were reported and Young's moduli of up to 106 GPa.

On the more academic side, Statton's original work was resumed with the difference that now UHMW-PE single crystals were employed. Kanamoto et al.[48] prepared single crystal mats of UHMW-PE which were extruded in the solid state at low EDR values, 6–25×, and drawn subsequently. Tensile moduli of up to 220 GPa were found at a corresponding draw ratio of about 250×. From the various studies on the drawing of single crystal mats from UHMW-PE,[49–51] the following main conclusions can be inferred:

(a)   Single crystal mats based on UHMW-PE are extremely ductile in a temperature range close to but below the melting temperature, as demonstrated by maximum draw ratios in the range 200–400×. However, the experimental observation that PE single crystals unfold quite easily in the direction of the applied stress at elevated temperatures had been found before by Statton.[22]

(b)   In the case of UHMW-PE single crystal mats, the combination of high molecular weight and high achievable draw ratios results in nearly perfect chain-extended polyethylene structures, as reflected by various measured properties compared to theoretical values for perfect crystals (see Table 3).

The use of UHMW-PE single crystals is of less practical interest in view of the low polymer concentrations in solutions which are a prerequisite for growing single crystals, typically below 0·05%, although the short-term mechanical properties obtained after super-drawing are superior. Apart from the impractical low solution concentrations, processing of precipitated single crystals by continuous extrusion is a slow and tedious

TABLE 3

COMPARISON OF VARIOUS (MAXIMUM) PROPERTIES MEASURED ON ULTRA-DRAWN UHMW-PE SINGLE CRYSTALS IN COMPARISON WITH THEORETICAL VALUES

| Property | Experimental | Theoretical |
|---|---|---|
| Modulus (GPa) | 220 | 250 (Ref. 7) |
| Density (kg m$^{-3}$) | 990 | 1 000 |
| Heat of fusion (cal g$^{-1}$) | 66·9 | 70 |
| Thermal expansion coefficient (°C)$^{-1}$ | $-1·16 \times 10^{-5}$ | $-1·18 \times 10^{-5}$ |

process since the temperature should be kept strictly below the melting point to avoid a dramatic loss in drawability of the extrudate upon melting and recrystallization; see Section 3.

These disadvantages on the points of extrusion and low solution concentrations were eliminated by the so-called 'gel-press-method'.[52] A semi-dilute solution of UHMW-PE was cooled and the resulting gel was pulverized to gel particles of about 80 $\mu$m. The gel particles were accumulated, compressed into sheets to be able to remove most of the solvent, and subsequently shaped via a heated die and post-drawn. During the die-drawing procedure, the solvent acts as a plasticizer and evaporates during post-drawing. The advantage of this process is that large diameter monofilaments or tapes can be produced based on UHMW-PE with tenacities above 2 GPa and corresponding moduli of up to 130 GPa.

A new approach which does away with the necessity of using a solvent to produce ultra-drawable precursors, such as single crystals mats or as-spun gel-filaments, was announced by Smith et al.[53] using so-called 'virgin' disentangled PE. With the use of special polymerization conditions, UHMW-PE films were produced showing ultra-draw characteristics. The as-polymerized films were drawn at 135°C to produce oriented structures possessing a tenacity of 3·5 GPa and corresponding moduli of 110 GPa.

## 3.    BASIC ASPECTS

### 3.1.    Mechanism for Ultra-drawing
At first sight it seems impossible to indicate a common mechanism for the various techniques and processes that have been developed for the production of high-modulus/high-strength polyethylene structures. However, confining considerations to the molecular scale, the prime observation has been that solution-spun/cast UHMW-PE, *even after complete removal of solvent*, is ultra-drawable as compared with melt-crystallized samples of identical chemical composition. This is, in fact, the essence of the invention of the gel-spinning process;[42] the constraints which limit the drawability of melt-crystallized UHMW-PE (see Fig. 4, Section 2) are removed by spinning/casting from semi-dilute solutions. The role of the solvent is to induce a favourable structure for ultra-drawing but once this structure is formed the solvent can be removed provided that the precursor thus generated is not destroyed by (partial) melting or dissolution (see below). From the influence of the initial polymer concentration in solution on the observed maximum drawability,

it was concluded that the strongly enhanced drawability upon spinning/casting from solution is due to a reduction of the entanglement-density in the gels or solid structures generated.[54] In dissolving UHMW-PE, the number of contacts between chains decreases proportionally to the degree of dilution, i.e. disentangling of chains. In the limiting case of dilute solutions ($\phi < \phi^*$), chains are separated and, upon cooling, individual folded-chain crystals will precipitate. Spinning or casting from solutions will yield folded-chain crystals during the quenching/solidification step as well, but a certain number of entanglements present in solution will be trapped in the gel or solid structure generated. Due to this trapping of entanglements coherent, gel-like (gel-spinning) filaments will be obtained upon spinning/quenching. In the case of hypothetical melt-spinning of UHMW-PE the situation in the melt (high entanglement density) is retained to a large extent in the solidified/quenched filaments. In the proposed model for the drawability of polyethylene in relation to solution concentration and crystallization history, it is assumed that the trapped entanglements act as physical cross-links that are semi-permanent on the time scale of the drawing experiment. In the case of melt-crystallized UHMW-PE the high entanglement density per chain is prohibitive to ultra-drawing whereas spinning/casting provides an optimum with respect to the entanglement density and coherence between the individual crystals to make in-line spinning/drawing feasible on a technological scale. Figure 10 shows the various possibilities with respect to the chain topology, taken from Reference 29. In Fig. 10 trapped entanglements are visualized as topologically interlocked loops. It should be noted that this model is a simplification of reality since the nature of entanglements is an experimentally elusive subject in macromolecular physics.[21] However, the fact is that the discovery of gel-spinning and its explanation in terms of removal of 'molecular constraints', interpreted as entanglements, through dissolution prior to drawing has proved to be a viable concept which has not been challenged up to now. Studies on ultra-drawing of single crystals based on UHMW-PE[48–50] demonstrate that, in the case of complete chain-disentangling through dilute solution, extremely high draw ratios can be obtained in the solid state in a temperature range close to but below the melting point, resulting in nearly perfect chain-extended PE structures.

Without specifying the nature and role of entanglements in the drawing process, it can be stated as a fact that chains are mutually uncrossable and that, to reach a situation in which chains are aligned parallel, the long-chain molecules of UHMW-PE have to be disentangled prior to tensile drawing. The formation of folded-chain type crystals as an intermediate

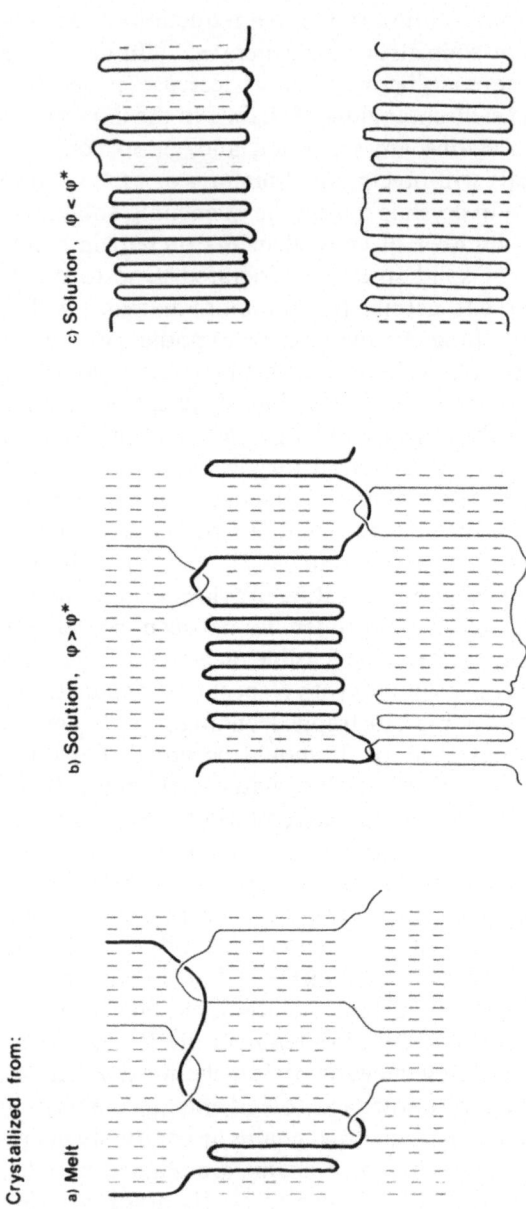

FIG. 10. Chain topology in crystallized UHMW-PE showing polymer chain (bold curve) with entangled (light curve) and non-entangled neighbours (dashed curve). (Reproduced from Reference 29, courtesy of Butterworths.)

stage does not present a problem in the case of polyethylene since these crystals are ductile and unfold in the direction of the applied stress.

### 3.2. Analysis of Other Techniques/Processes

As mentioned in Section 2, porosity has been claimed to be important in ultra-drawing of UHMW-PE, on the basis of free-volume considerations. However, one should not confuse intermolecular (microscopic) free volume with intercrystalline (macroscopic) voids as induced by extraction of solvent from spun/cast filaments or films.[55] This is demonstrated convincingly by the studies on drawing of UHMW-PE single crystals. Compression at room temperature to promote toughness in UHMW-PE single crystal mats[56] or even solid-state extrusion at high pressure does not affect the ultra-draw characteristics, despite the fact that 'porosity' has been removed almost completely.

Melt-kneading of UHMW-PE and paraffin wax is a seemingly different process involving two solid constituents, at least at room temperature. The extrusion/spinning of this mixture is performed at temperatures between 180 and 300°C. For the system UHMW-PE/paraffin wax, the thermodynamic behaviour can be calculated straightforwardly. The equilibrium melting (dissolution) temperatures are derived from the well known Flory melting-point depression relationship:[57]

$$\frac{1}{T_m} - \frac{1}{T_m^\circ} = \left(\frac{R}{\Delta H_u}\right)\left(\frac{V_u}{V_1}\right)[(1 - \phi) - \chi(1 - \phi)^2] \tag{5}$$

In eqn (5), $T_m$ and $T_m^\circ$ are the dissolution and the melting temperature of PE, $\Delta H_u$ is the heat of fusion per repeating unit, $V_u/V_1$ the molar volume ratio and $\phi$ the polymer volume fraction. Figure 11 shows the calculated dissolution temperatures for two molecular weights of paraffin and two values of the interaction parameter $\chi$. As is clear from Fig. 11, the actual extrusion temperature ($>180°C$) used in the melt-kneading process is far above the dissolution temperature, and consequently the paraffin/UHMW-PE system is a true solution during spinning/extrusion. Consequently melt-kneading is identical with solution-spinning/gel-spinning employing volatile solvents.

All processes discussed up to now involve a dissolution stage to disentangle the long-chain PE molecules either partly, to create a loose entanglement network structure as the ideal precursor for ultra-drawing (gel-spinning), or completely, i.e. production of UHMW-PE single crystals. In retrospect this concept has been used to explain the phenomenon of

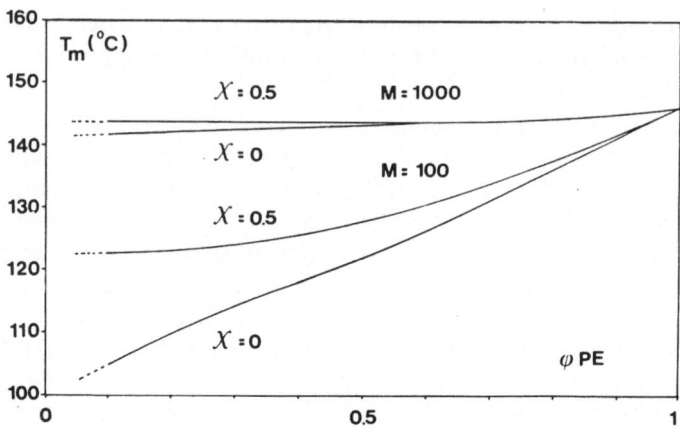

FIG. 11. UHMW-PE liquidus calculated for system UHMW-PE/paraffin ($M = 100$ and $M = 1000\,kg\,kmol^{-1}$); the actual values for the interaction parameter, $\chi$, will be between the two chosen values $\chi = 0\cdot5$ and $\chi = 0$.

surface growth in terms of a loose entanglement network on the rotor surface.[36,37]

However, as indicated by Smith et al.,[53] UHMW-PE films exhibiting ultra-draw characteristics can be produced by direct polymerization. It is also well known that nascent PE or PP reactor powders, obtained by heterogeneous polymerization, are to some extent disentangled since the chains grow initially independently and form folded or extended-chain type crystals depending on the polymerization conditions.[58] Therefore a polymerization process could be envisaged producing disentangled UHMW-PE powder directly in the reactor. If so, the 'only' problem left towards a solvent-free route is to process disentangled UHMW-PE powder which is in fact healing of the particles so as to form continuous structures without destroying ultra-drawability. Literature data indicate that crystallization and processing memories persist in polyethylene melts for many hours.[59,60] In view of the high molecular weight involved ($M_w$ typically above $10^6\,kg\,kmol^{-1}$) one expects a 'temperature-time window' for processing (extrusion/spinning) where the initially disentangled status is not completely destroyed with loss of drawability. However in the authors' experience such a '$T$–$t$ window' does not exist in practice, for two related reasons:

(a)   The rheological properties of UHMW-PE melts, at least in shear and within time-scales (frequency ranges) of practical interest, are independent of the previous sample history. Figure 12 shows the

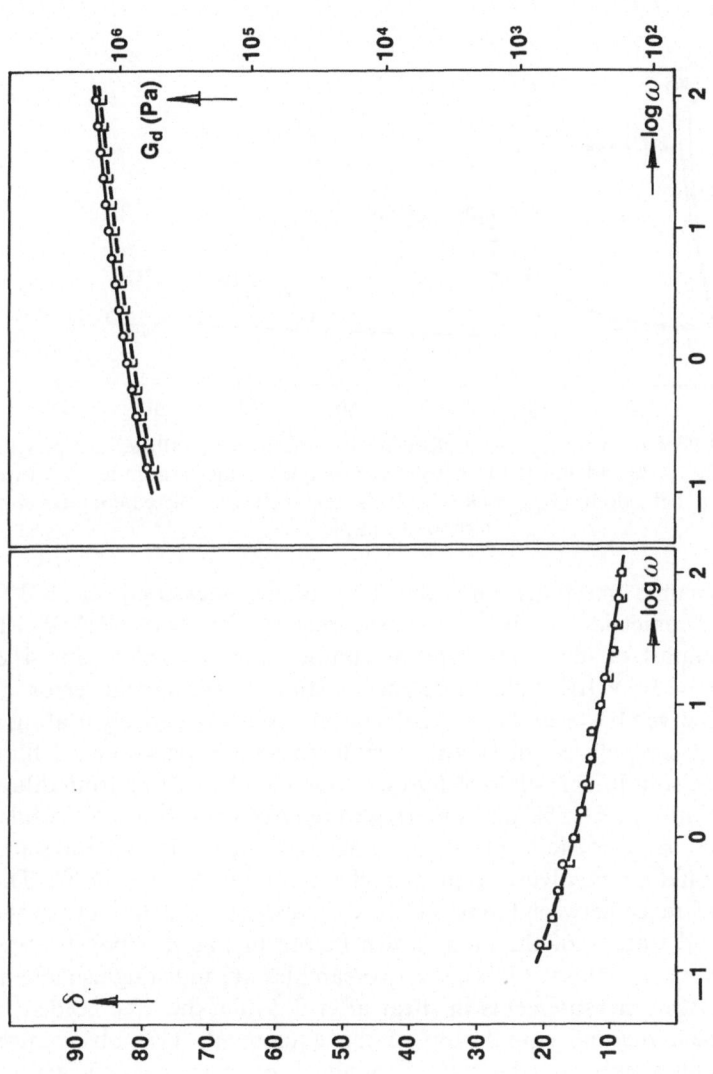

FIG. 12.   Dynamic modulus, $G_d$, and loss angle, $\delta$, measured at 160°C (Rheometrics mechanical spectrometer): L, solution-cast; ○, melt-crystallized/compression-moulded UHMW-PE sample.

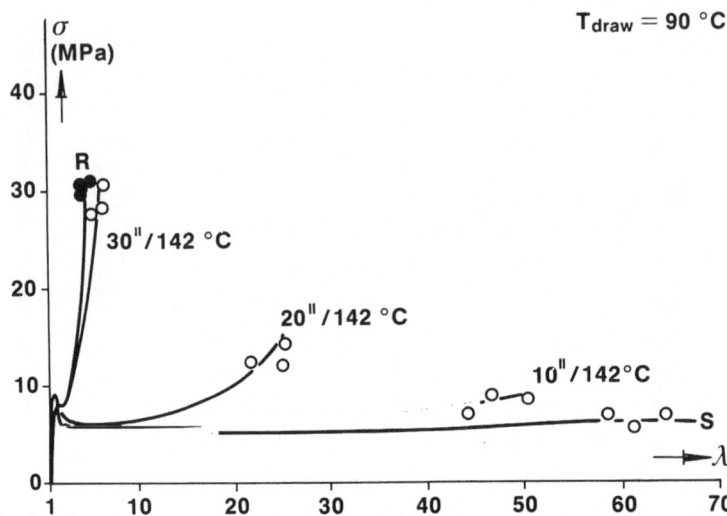

FIG. 13.    Effect of melting/recrystallization of solution-crystallized samples (S) (respectively 10, 20 and 30 s at 142°C followed by quenching to room temperature) on drawing behaviour ($T_{draw} = 90°C$); R is for melt-crystallized/compression-moulded sample.

dynamic modulus and the loss angle, measured at 160°C (Rheometrics mechanical spectrometer), for two UHMW-PE samples of identical chemical composition (Hostalen Gur 412, $M_w = 1.5 \times 10^6$). No difference within experimental error is observed between the two melts obtained by respectively heating a melt-crystallized sheet and a sintered solution-crystallized film. The solution-crystallized film was obtained by casting from dilute solutions ($<0.5\%$) and sintering to remove voids below $T_m$, whilst the melt-crystallized sheet was obtained via compression moulding involving a press-cycle of about 1 h at 200°C. The difference between the two films with respect to drawability in the solid state is of the kind shown before in Fig. 8. Upon heating above $T_m$, about 135°C, the two samples are indistinguishable at least in measurements in shear after 3–5 min, the time needed to reach thermal equilibrium in the equipment. This observation implies that in the actual practice of extrusion (shear) no advantage is gained by starting from disentangled UHMW-PE powder or related structures (gel particles etc.).

(b)    If disentangled UHMW-PE, in the form of cast-films or single

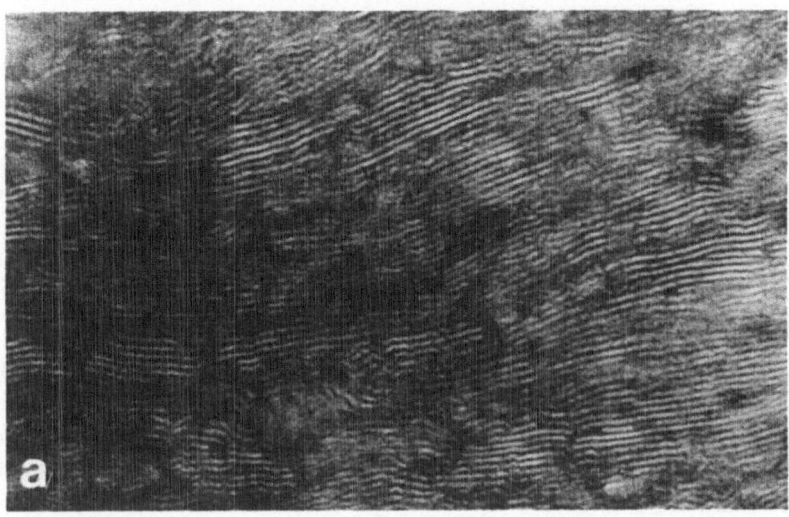

**100nm**

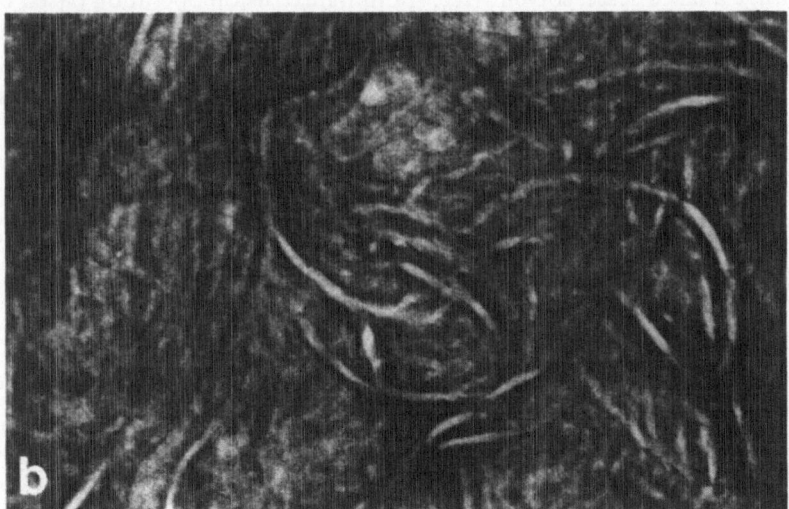

FIG. 14. Morphology of (a) solution-crystallized (cast from 1·5% solution in decalin) UHMW-PE and (b) recrystallized sample (heated to 150°C for 60 s and quenched).

crystal mats, is heated above $T_m$ and recrystallized, the ultra-draw characteristics are almost lost immediately. Figure 13 shows the effect of heating a solution-cast film (S) above $T_m$ (135°C), in this case 142°C, and recrystallizing by quenching to room temperature. Heating for less than 1 min in the melt is sufficient to cause complete loss of the ultra-draw characteristics, and the sample is indistinguishable in drawing behaviour from a melt-crystallized/ compression-moulded sheet (R). Figure 14 shows the corresponding change in morphology upon heating for about 1 min above $T_m$ followed by quenching to room temperature.

The short time scale for destroying completely the ultra-draw characteristics upon recrystallization of disentangled UHMW-PE structures reflects a fast molecular reorganization upon melting. The question as to whether re-entangling (reptation[29]) or only local, short-range motions are involved is outside the scope of this review. The main conclusion in practice is that the operating temperature during processing of disentangled UHMW-PE should be kept below the melting temperature, i.e. processing should be restricted to solid-state extrusion. In view of the poor flow properties of solid PE particles it is difficult to envisage a large-scale, solvent-free, spinning/extrusion operation based on disentangled UHMW-PE. However, for the production of large diameter monofilaments or PE tape, solid-state extrusion of disentangled UHMW-PE or related processes such as gel-pressing[52] seem technologically feasible.

In conclusion it can be said that gel-spinning of UHMW-PE is, at the time of writing, the optimum route for continuous in-line *spinning* (low viscosity due to presence of solvent)/*drawing* (optimum morphology, ultra-draw characteristics) in the production of high-strength/high-modulus PE fibres.


## 4. SOME PROPERTIES OF HIGH-PERFORMANCE PE FIBRES

The properties presented in this section were obtained from HP-PE fibres originating from the joint development programme of Toyobo and DSM. One should realize that these properties are of a dynamic nature due to a continuous effort being made towards product and process optimization. A selection has been made to show the potential of HP-PE fibres in some promising applications.

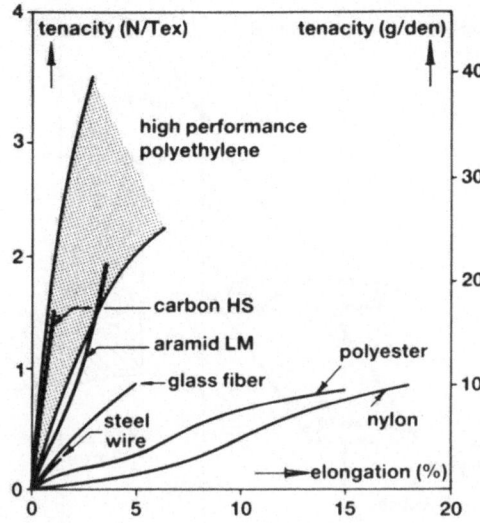

FIG. 15. Stress–strain behaviour of various yarns; 10% min$^{-1}$, 23°C.

### 4.1. Stress–Strain Behaviour

Figure 15 shows tensile strengths ranging from 25 to 40 g den$^{-1}$ at a corresponding strain at break of 6–3% (measured at 23°C/crosshead speed 10% min$^{-1}$). Young's moduli ranging from 600 to 1800 g den$^{-1}$ were recorded, which implies that the specific modulus even of high-strength carbon is surpassed. To avoid any confusion about textile units, a comparison for several materials is shown in Table 4. A comparison as to specific strength/modulus between high-performance PE and some well

TABLE 4
COMPARISON OF TEXTILE UNITS

| Material | Density (g cm$^{-3}$) | GPa | N tex$^{-1}$ | g den$^{-1}$ |
|---|---|---|---|---|
| Nylon | 1·14 | 1·0 | 0·88 | 9·94 |
| Polyester | 1·38 | 1·0 | 0·73 | 8·21 |
| PP | 0·91 | 1·0 | 1·10 | 12·45 |
| Aramids | 1·45 | 1·0 | 0·69 | 7·81 |
| Carbon | 1·77 | 1·0 | 0·56 | 6·40 |
| Glass | 2·5 | 1·0 | 0·40 | 4·53 |
| Steel | 7·8 | 1·0 | 0·13 | 1·45 |
| HP-PE | 0·98 | 1·0 | 1·02 | 11·56 |

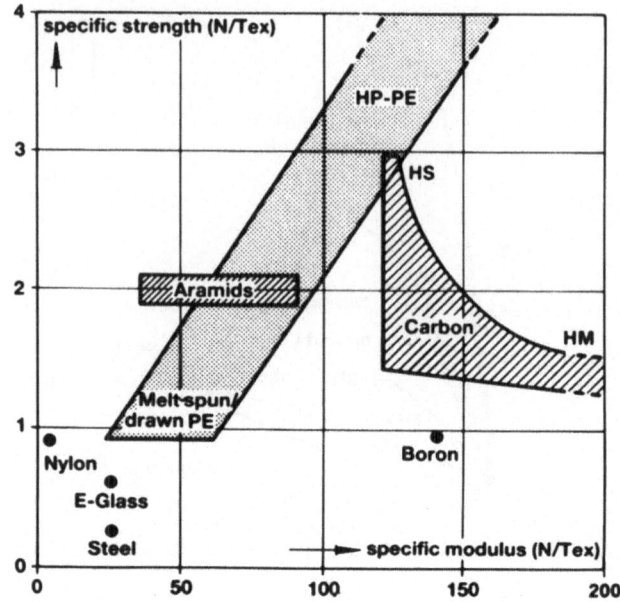

FIG. 16.    Specific strength vs. specific modulus.

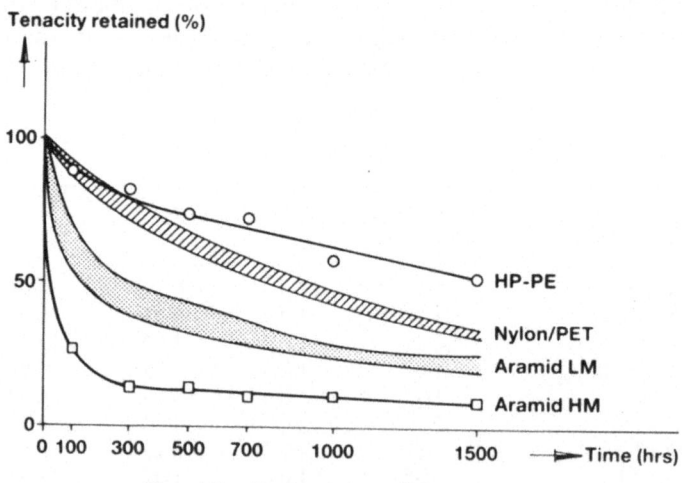

FIG. 17.    Resistance to light.

known inorganic/organic fibres is shown in Fig. 16. Apart from having the advantage of extreme high strength/stiffness, the textile engineer is able to match the fibre properties to the requirements of the end-user by changing only a few process parameters during the spinning/drawing operation, such as draw ratio and concentration.

## 4.2. Light Resistance
The strength retention after exposure to (artificial) sunlight (JIS L1013-7.18) in comparison with aramid fibres is given in Fig. 17; it is found to be superior. The exposure time of 300 h in the Fadometer corresponds roughly to an outdoor exposure time of 6 months. For various marine applications (long exposure times), the advantage is clear.

## 4.3. Chemical Resistance
Due to the simple chemical structure of polyethylene with C–C and C–H bonds only, no chemical groups are available for attack in harsh environments. The effect of extremely low pH ($H_2SO_4$, 95%) and high pH (NaOH, 50%) on the strength levels in comparison with those of aramid fibres is shown in Fig. 18. The favourable behaviour of HP-PE could be advantageous in filtering applications in the chemical industry.

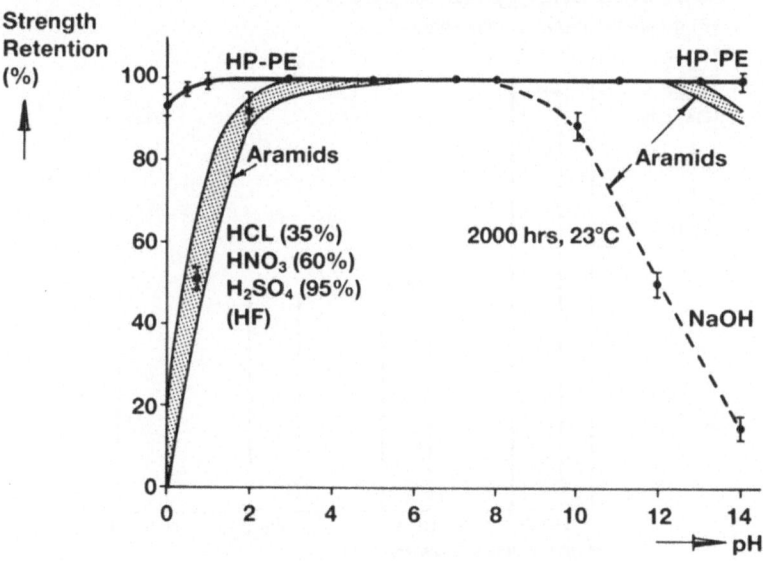

FIG. 18.   Chemical resistance; 120 h, 23°C.

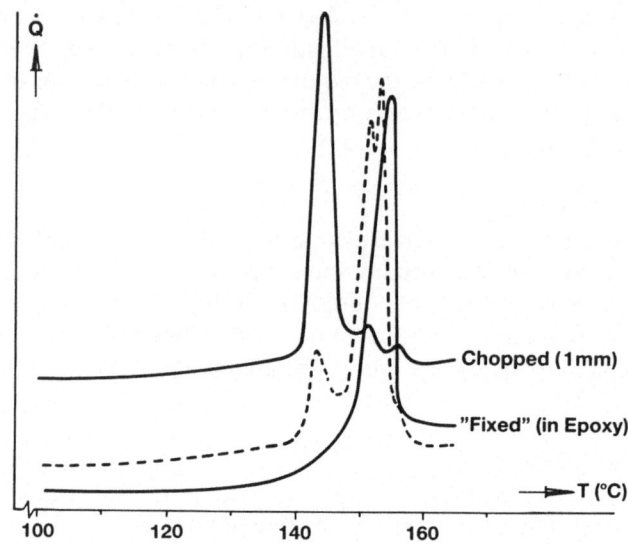

FIG. 19.   Melting behaviour of HP-PE. (Reproduced from Reference 29, courtesy of Butterworths.)

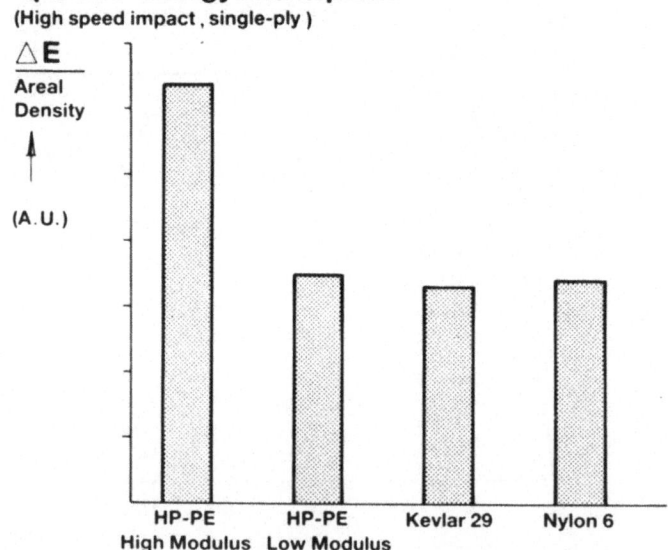

FIG. 20.   Specific energy absorption of various fibres/fabrics.

## 4.4. Melting Behaviour

The melting behaviour of HP-PE fibres seems at first sight rather complex (see the dashed line in Fig. 19) due to various constraints imposed on the fibre during the measurement. However, this complex pattern is in fact due to a superposition of two melting modes:

(a) Non-constrained melting (e.g. chopped fibres in silicone oil), showing a single peak at 144°C (5°C min$^{-1}$).

(b) Fully constrained melting (e.g. fibre embedded in an epoxy matrix), showing a single endotherm at about 155°C (5°C min$^{-1}$).

The higher melting point of constrained PE fibres makes it even possible to produce fully HP-PE/PE based composites which offer possibilities in ballistic applications such as 'hard armour'.[61,62]

## 4.5. Impact Resistance

A combination of high modulus, high breaking strength and an elongation at break of 3–6% leads to a high (calculated) value of the work to break of the fibre (see Table 5). A comparison of two types of HP-PE, Kevlar-29 and nylon-6, is given in Fig. 20, with the absorbed energy measured in high-speed impact testing. It is in particular this superior impact property that makes HP-PE a good candidate for several protective textile and composite applications.[62]

## 4.6. Abrasion/Flex/Bending Properties

Due to the low friction coefficient of polyethylene (UHMW-PE is famous for this unique property as an engineering plastic), the abrasion resistance,

TABLE 5
MODULUS AND WORK-TO-BREAK FOR HIGH-PERFORMANCE FIBRES

| Material | Young's modulus ($N$ $tex^{-1}$) | Work to break[a] ($N$ $m$ $tex^{-1}$) $\times 10^3$ |
|---|---|---|
| Carbon   HS | 125 | 1–2 |
| Aramid   HM | 80 | 2·5–3·0 |
| Aramid   LM | 40 | 3·5–4·0 |
| Polyethylene | 50 | 11–13 |
| Polyethylene | 80 | 8–10 |
| Polyethylene | 100 | 8–9 |
| Polyethylene | 130 | ~8 |

[a] Calculated by integration of the stress–strain curve.

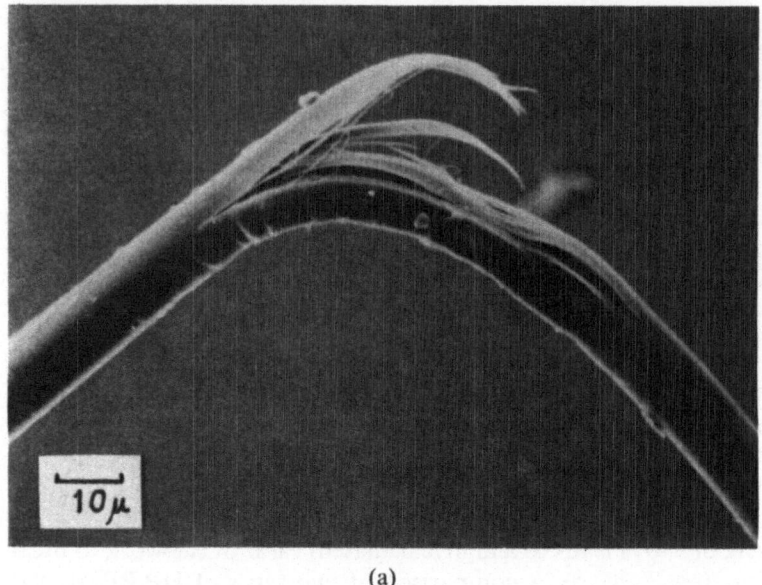

(a)

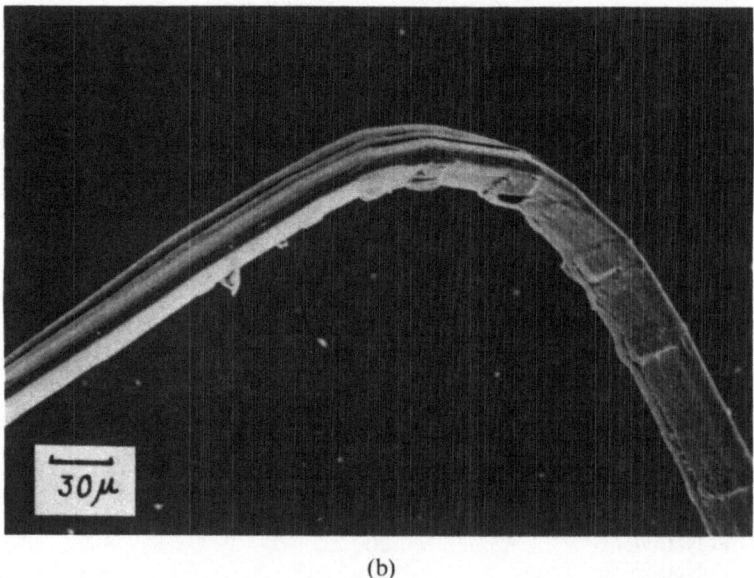

(b)

FIG. 21.   Behaviour of fibres in loop-strength tests. (a) Kevlar 49; (b) HP-PE.

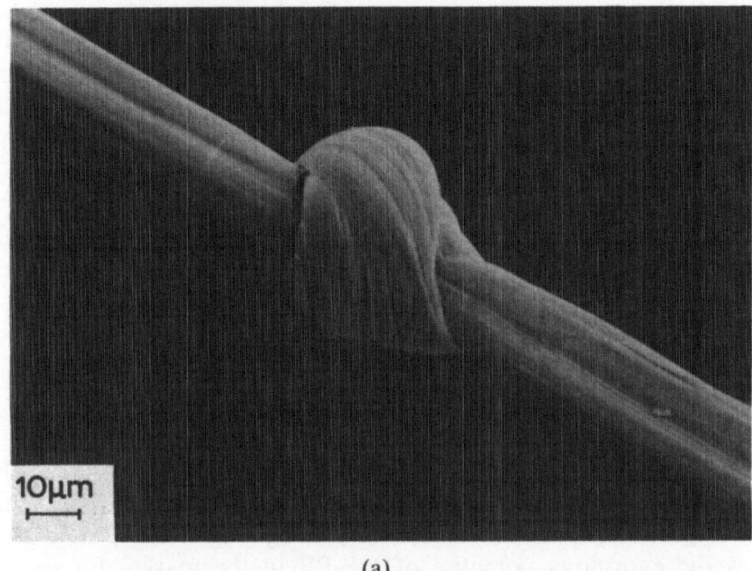

(a)

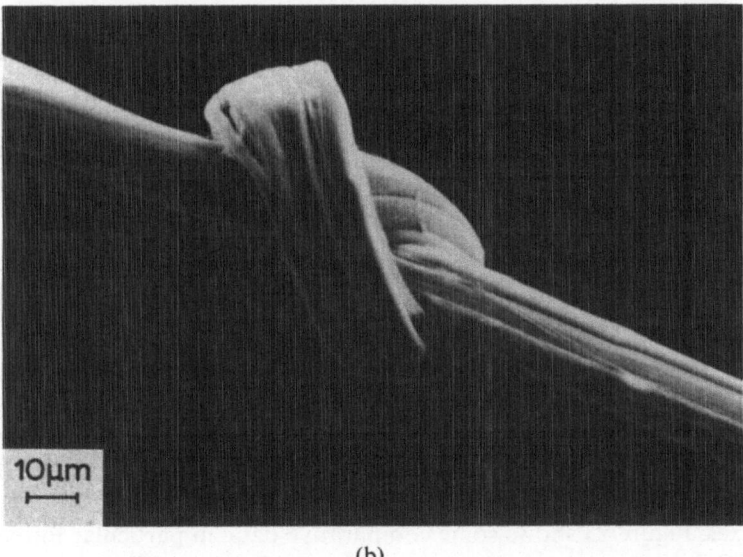

(b)

FIG. 22. SEM micrographs revealing knots in monofilaments of (a) HP-PE and (b) aramids.

### TABLE 6
YARN PROPERTIES

|  | HP-PE | Kevlar 29 | Kevlar 49 | Carbon HT | Carbon HM |
|---|---|---|---|---|---|
| Abrasion resistance (cycles to break) | $>110 \times 10^3$ | $9.5 \times 10^3$ | $5.7 \times 10^3$ | 20 | 120 |
| Flex life (cycles to break) | $>240 \times 10^3$ | $37 \times 10^3$ | $43 \times 10^3$ | 5 | 2 |
| Knot strength (g den$^{-1}$) | 10–15 | 6–7 | 6–7 | 0 | 0 |
| Loop strength (g den$^{-1}$) | 12–18 | 10–12 | 10–12 | 0.7 | 0.1 |

measured according to JIS L1095-7.10.2, is absolutely superior to that of other high-performance fibres, as shown in Table 6. More generally speaking, the properties of the fibre during single or repeated bending show the enormous potential of HP-PE in the market for ropes and lightweight fabrics. Both knot strength and loop strength are the highest found in any man-made material related to the intrinsic flexible nature of polyethylene macromolecules. This is also demonstrated in Fig. 21, showing the aramid and HP-PE fibres after being tested for loop strength. Before rupture the tested monofilaments were examined in the scanning microscope, revealing kinking in the case of PE and splitting for the aramids. The favourable bending behaviour of PE is shown once more in Fig. 22, revealing knotted monofilaments.

### 4.7. Long-term Properties

Creep of HP-PE fibres is dependent on a large variety of factors, such as $E$ modulus, molecular weight and molecular weight distribution, spinning concentration etc., and is subject to change in the future due to improved control of process and product quality and optimization with respect to the aforementioned parameters. Therefore no definite creep curves for HP-PE can be given. However, it remains an interesting question whether there will be a consistent difference between HP-PE fibres produced from low-volatility solvents and fibres from paraffinic oil, in particular paraffin waxes. Figure 23 shows some comparative data, in particular for creep at 80°C, i.e. above the melting point of paraffinic wax. These data suggest that residual solvent could promote creep, which is understandable in terms of lubrication of individual microfibrils.

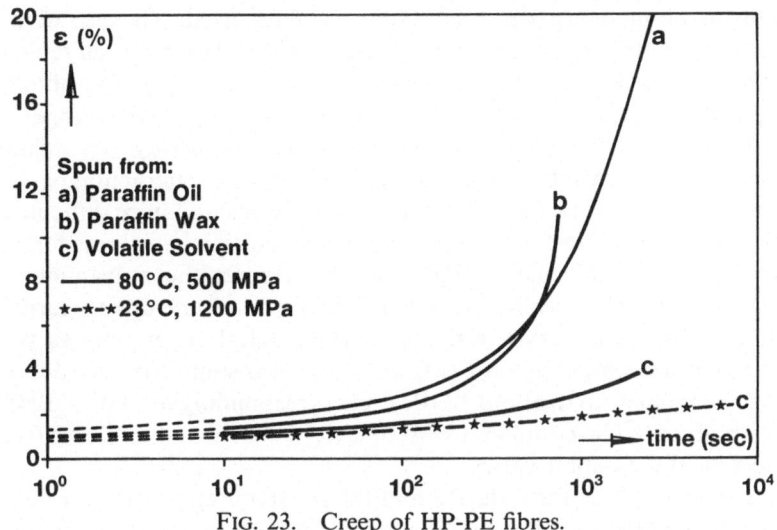

FIG. 23. Creep of HP-PE fibres.

## 5. OTHER FLEXIBLE POLYMERS

Since the invention of the gel-spinning process for polyolefins, a lot of effort has been devoted towards the utilization of this concept for other flexible polymers.

High molecular weight polypropylene was investigated in detail by Peguy and Manley;[63] by ultra-drawing of dried gels oriented PP structures were obtained possessing moduli of up to 36 GPa and corresponding tenacities of about 1 GPa. Their experimentally obtained moduli are close to the theoretical limits (compare with Table 1), but of course rather low in comparison with the values for HP-PE due to the helical conformation of isotactic PP in the crystalline state.

An interesting candidate is polyvinylalcohol (PVAL). Despite its atactic character, PVAL is crystallizable, possesses the all-*trans* conformation in the crystalline state, and the crystal structure is very similar to that of polyethylene. The advantage over PE is its higher melting point, > 200°C. Allied researchers copied their 'xero-gel' process[44] for the production of PVAL fibres.[64] High molecular weight PVAL was obtained by low-temperature photo-initiated polymerization of vinyl acetate followed by methanolysis. Molecular weights above $5 \times 10^5$ kg kmol$^{-1}$ ($M_w$) and up to several millions were used in their spinning experiments. Tenacities of the fibres of up to about 20 g den$^{-1}$ (2·3 GPa) and corresponding tensile moduli of about $600 \, \text{g den}^{-1}$ (70 GPa) were reported. The use of high molecular weight was claimed to be essential.[64,65] However, as reported by Toray,[66] similar values for tenacity and modulus can be obtained using rather standard molecular weight PVAL ($M_w$ in the range 1·3–1·8 × kmol$^{-1}$). This shows immediately the difficulty in relating in particular the tenacity to the chain length. Stronger interaction between chains makes the drawing process, i.e. chain-extension, more difficult but renders a higher value for tenacity and in particular a better resistance to creep for lower molecular weights. In this respect the polyamides are attractive due to the specific hydrogen bonding in the crystalline state, but no impressive values have been obtained via solution-spinning as yet.[67] High molecular weight nylon-6 ($M_w$ 3·5 × 10$^6$) was spun from solution and drawn. Tenacities up to about 1 GPa and corresponding moduli of 19 GPa were obtained, to be compared with tenacity 0·8 GPa and modulus 6 GPa for technical melt-spun yarns.

A final example of the utilization of the gel-spinning process for flexible polymers is poly(acrylonitrile). Technical PAN yarns based on standard molecular weights ($M_w$ 5–10 × 10$^4$ kg kmol$^{-1}$) possess tensile strengths of 0·8 GPa and moduli of about 15 GPa.[68] High molecular weight PAN ($M_w$ about 10$^6$ kg kmol$^{-1}$) was employed[69] but no significant increase in properties was obtained, at least with respect to tenacity, viz. 0·8 GPa and corresponding moduli of about 27 GPa. A major improvement in tenacity was reported[70] by spinning of high molecular weight PAN from solutions which contain bivalent metal ions, for example $Zn^{2+}$. The presence of bivalent ions promotes the formation of homogeneous gel-filaments and prevents to some extent the L–L phase separation which usually occurs during coagulation in conventional wet-spinning processes. The combination of high molecular weight and specific metal ions results in tenacity values greater than 1·5 GPa.[70] Similar values have been reported by Japan Exlan.[71] An intriguing question which remains is whether these HS-PAN fibres could lead to better carbon fibres with respect to tenacity and improved toughness. In other words, will a higher tenacity, due to improved chain orientation and higher molecular weight, of the precursor be reflected in the properties of the carbon fibres after oxidation/carbonization?[72]

## 6. CONCLUSIONS

The properties of the various fibres based on flexible macromolecules, as presented in this review, are of a dynamic nature; due to the continuous efforts made at process and product optimization, improvements are foreseen.

In the case of HP-PE fibres one could classify three application areas:

New, as yet unexplored markets related to the unique properties of HP-PE.

Combination of HP-PE with other high-performance-fibres in hybrid structures, both composites and textiles.

Substitution of currently used high-performance-fibres, such as aramids, carbon and glass, mainly depending on price/performance criteria.

In general, apart from gel-spinning (solution-spinning of high molecular weight flexible polymers followed by ultra-drawing), increased activity is foreseen towards the utilization of 'gel technology' for the production of two- and three-dimensional strong and stiff structures based on flexible polymers.

## REFERENCES

1. CAROTHERS, W. and HILL, J. W. (1932). *J. Am. Chem. Soc.*, **54**, 1586.
2. W-FANG HWANG (1985). Proc. Symp. *Formation, structures and properties of high-modulus and high-tenacity fibres*, Kyoto, p. 23.
3. W-FANG HWANG (1985). *Proc. Int. Symp. Fibre Sci. Technol. (ISF)*, Hakone, p. 39.
4. SMITH, P., MATHESON, R. R. and IRVINE, P. A. (1984). *Polymer Commun.*, **25**, 294.
5. TANNER, D., FITZGERALD, J. A., KNOFF, W. F. and PIGLIACAMPI, J. J. (1985). *Proc. ISF*, Hakone, p. 31.
6. PEREPELKIN, K. E. (1972). *Angew. Makromol. Chem.*, **22**, 181.
7. SAKURADA, I. and KAJI, K. (1970). *J. Polym. Sci. C*, **31**, 57.
8. STORKS, K. H. (1938). *J. Am. Chem. Soc.*, **60**, 1753.
9. JACCODINE, R. (1955). *Nature*, **176**, 305. KELLER, A. (1957). *Phil. Mag.*, **2**, 1171. FISCHER, E. W. (1957). *Z. Naturforsch.*, **12a**, 753. TILL, P. H. (1957). *J. Polym. Sci.*, **24**, 301.
10. MARK, H. F. (1971). In: *Polymer science and materials* (Ed. H. F. Mark and A. V. Tobolsky), Wiley-Interscience, New York, p. 236.
11. BOUDREAUX, D. S. (1973). *J. Polym. Sci. Phys. Ed.*, **11**, 1285.
12. PETERLIN, A. (1979). In: *Ultra-high modulus polymers* (Eds. A. Ciferri and I. M. Ward), Applied Science Publishers, London, p. 279.
13. JURGENLEIT, W. (1962). US Patent 3,048,465 (appl. Germany, June 16, 1956).
14. SATO, H. and HIRAI, N. (1962). Jap. Sho 37-9765.
15. ZWICK, M. (1965). NL 6501248.
16. BLADES, H. and WHITE, J. R. (1963). US Patent 3,081,519.
17. MITSUHASHI, S. (1963). *Bull. Text. Res. Inst.*, **66**, 1.

18. PETRIE, C. J. S. (1979). *Elongation flows*, Pitman, London, Ch. 4.
19. MACKLEY, M. R. and SAPSFORD, G. S. (1982). In: *Developments in oriented polymers* (Ed. I. M. Ward), Applied Science Publishers, London.
20. MACKLEY, M. R. and KELLER, A. (1975). *Phil. Trans. Roy. Soc.*, **A278**, 29.
21. KELLER, A. and ODELL, J. A. (1985). *Colloid. Polym. Sci.*, **263**, 181.
22. STATTON, W. O. (1967). *J. Appl. Phys.*, **38**, 4149.
23. ISHIKAWA, K., MIYASAKA, K. and MAEDA, M. (1969). *J. Polym. Sci. A-2*, **7**, 2029.
24. MAEDA, M., MIYASAKA, K. and ISHIKAWA, K. (1970). *J. Polym. Sci. A-2*, **8**, 355.
25. BARHAM, P. J. and KELLER, A. (1976). *J. Mater. Sci.*, **11**, 27.
26. ANDREWS, J. M. and WARD, I. M. (1970). *J. Mater. Sci.*, **5**, 411.
27. CAPACCIO, G., GIBSON, A. G. and WARD, I. M. (1979). In: *Ultra-high modulus polymers* (Eds. A. Ciferri and I. M. Ward), Applied Science Publishers, London, Ch. 1.
28. STAMM, M., FISCHER, E. W., DETTENMAIER, M. and CONVERT, P. (1979). *Faraday Discuss. Roy. Soc. Chem.*, **68**, 263.
29. LEMSTRA, P. J. and KIRSCHBAUM, R. (1985). *Polymer*, **26**, 1372.
30. SADLER, D. M. and ODELL, J. A. (1980). *Polymer*, **21**, 479.
31. (1984). *Eur. Chem. News*, **43**, 22.
32. GIBSON, A. G., WARD, I. M., COLE, B. N. and PARSONS, B. (1974). *J. Mater. Sci.*, **9**, 1193.
33. PERKINS, W. G., CAPIATI, N. J. and PORTER, R. S. (1976). *Polym. Eng. Sci.*, **16**, 3.
34. NAT. RES. DEV. CORP., European Patent Appl. 38798.
35. PENNINGS, A. J., VAN DEN MARK, J. M. A. A. and BOOIJ, H. C. (1970). *Colloid Polym. Sci.*, **236**, 99.
36. KELLER, A. AND BARHAM, P. J. (1981). *Plastics Rubber Int.*, **6**, 19.
37. BARHAM, P. J. and KELLER, A. (1985). *J. Mater. Sci.*, **20**, 2281.
38. PENNINGS, A. J. (1977). *J. Polym. Sci. Polym. Symp.*, **59**, 55.
39. ZWIJNENBURG, A. and PENNINGS, A. J. (1975). *Colloid Polym. Sci.*, **253**, 452.
40. DSM/STAMICARBON, NL 7605370 (1976)/US Patent 4,137,394.
41. ZWIJNENBURG, A. (1978). Ph.D. Thesis, University of Groningen, Netherlands.
42. DSM/STAMICARBON, NL 7900990 (1979); 7904990 (1979)/US Patent 4,344,908; 4,422,993; 4,436,689.
43. PREVORSEK, D. C. (1982). *Polymer liquid crystals*, Academic Press, p. 357.
44. ALLIED FIBRES, Eur. Patent 64167 (1982).
45. KALB, B. and PENNINGS, A. J. (1980). *Polymer*, **21**, 3.
46. WILDING, M. A. and WARD, I. M. (1978). *Polymer*, **19**, 969.
47. MITSUI, Eur. Patent 115.192 (1984).
48. KANAMOTO, T., TSURUTA, A., TANAKA, K., TAKEDA, M. and PORTER, R. S. (1983). *Polymer. J.*, **15**, 327.
49. MIYASAKA, K., in Reference 3, p. 27.
50. MATSUO, M. and SAWATARI, V., in Reference 3, p. 152.
51. KANAMOTO, T., TANAKA, K., TAKEDA, M. and PORTER, R. S., in Reference 3, p. 155.
52. TOYOBO, Eur. Patent 84113352.3/141418.
53. SMITH, P., CHANZY, H. D. and ROTZINGER, B. P. (1985). *Polym. Commun.*, **26**, 257.
54. SMITH, P., LEMSTRA, P. J. and BOOY, H. C. (1981). *J. Polym. Sci. Polym. Phys. Ed.*, **19**, 877.

55. LEMSTRA, P. J. and SMITH, P. (1980). *Brit. Polym. J.*, **12**, 212.
56. FURUHATA, K., YOKOKAWA, T. and MIYASAKA, K. (1984). *J. Polym. Sci. Polym. Phys. Ed.*, **22**, 133.
57. FLORY, P. J. (1953). *Principles of polymer chemistry*, Cornell University Press, Ch. XIII.
58. BOOR, J. (1979). *Ziegler-Natta catalysts and polymerizations*, Academic Press, New York, Ch. IV.
59. FOLLAND, R. and CHARLESBY, A. (1979). *Eur. Polym. J.*, **15**, 953.
60. SCHREIBER, H. P. (1983). *Polym. Eng. Sci.*, **23**, 422.
61. ALLIED CORP, Eur. Patent 110.021 (1984).
62. ADAMS, D. F., ZIMMERMAN, R. S. and WON CHANG, H. (1985). *SAMPE J.*, 44.
63. PEGUY, A. and MANLEY, R. ST J. (1984). *Polym. Commun.*, **25**, 39.
64. ALLIED FIBRES, Eur. Patent. 105.169 (1984).
65. PREVORSEK, D. C., in Reference 2, p. 20.
66. TORAY INDUSTRIES, Eur. Patent 146.084 (1985).
67. GOGOLEWSKI, S. and PENNINGS, A. J. (1984). *Polymer*, **26**, 1394.
68. HOECHST, Technical Bulletin 198b (Dolan-10).
69. ALLIED FIBRES, Eur. Patent 144.793 (1985).
70. DSM/STAMICARBON, Eur. Patent 114.983 (1984).
71. JAPAN EXLAN CO., US Patent 4,535,027 (1985).
72. SANTAPPA, M. (1982). *J. Indian Chem.*, **59**, 321.
73. LEMSTRA, P. J., VAN AERLE, N. A. M. J. and BASTIAANSEN, C. W. N. (1987). *Polymer J.*, **19**, 85.

*Note added on proof*

Since the time of writing this chapter considerable progress has been made in the area of UHMW-PE gel-spinning. For further reading see for example ref. 73.

# Chapter 3

# SPINNING FROM LYOTROPIC AND THERMOTROPIC LIQUID CRYSTALLINE SYSTEMS

A. CIFERRI

*Istituto di Chimica Industriale, University of Genoa, Genoa, Italy*

## 1. THEORETICAL BACKGROUND

Amorphous orientation with conventional flexible polymers is hindered by the natural tendency to recoil of the molecules, by the lack of a microscopic degree of order, and the occurrence of chain entanglement. Rigid polymers in a nematic mesophase should be more simply oriented than flexible ones since they have no tendency to recoil and exhibit a microscopic order of molecules described by the so-called order parameter. In fact, in order to fully orient the rigid nematic polymers, it is necessary to

(1)  increase the order parameter:[1]

$$S = \tfrac{1}{2}\langle(3\cos^2\theta - 1)\rangle$$

where $\theta$ is the average angle between the molecular axis and the director (a vector representing the average molecular orientation at any point of the liquid[1]) toward the value $S = 1$ expected for complete orientation;

(2)  orient the director along the flow direction or, according to an alternative (less precise[1]) description, to transform the 'poly-domain' texture (rich in disclinations) into a 'monodomain' texture.[2] The overall process is schematized in Fig. 1.

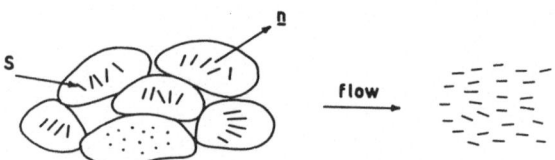

FIG. 1.   Increased order parameter and director orientation during spinning.

In the case of semi-rigid polymers which may form mesophases at rest, or only under the action of a flow field, one may expect that the orientation process exhibits characteristics typical of rigid mesogens coupled with those typical of flexible polymers.

In this first section we review recent results which pertain to mesophase formation: chain rigidity, order parameter, domain size and stability. In the second section we describe spinning data for four classes of polymers and attempt a correlation (largely empirical at the present time) with the data discussed in the first part. The four systems considered include chains with rigid conformations studied as lyotropic or thermotropic systems, i.e. poly(p-benzamide) (PBA), poly(p-phenylene terephthalamide) (PPTA) and poly(p-phenylene terephthalate), and chains with semi-rigid conformation again studied as lyotropic or thermotropic systems (e.g. poly-terephthalamide of p-aminobenzhydrazide, cellulose, and polyesters containing flexible methylene units). The emphasis is on the role of the mesophase and not on fiber structure, or the optimization of properties.

### 1.1. Chain Rigidity
The appearance of the mesophase begins at a critical polymer volume fraction, $v_2'$, and is completed at a slightly larger value, $v_2''$.[3] The most reliable method to determine the width of the biphasic gap is the extrapolation of the volume fraction of the isotropic phase, $\phi$, vs. the overall $v_2$ at $\phi = 1$ and $\phi = 0$.[4] $v_2'$ is inversely related to the axial ratio, $x$, of the molecule by a relationship which satisfactorily describes the experimental behavior:[3]

$$v_2' = \frac{8}{x}\left(1 - \frac{2}{x}\right) \tag{1}$$

Eqn (1) applies even to semi-rigid chains, but the axial ratio should then refer to the length of the Kuhn segment.[3] The theory leading to eqn (1) predicts that, for pure polymer ($v_2 = 1$), $x = 6.4$ is the minimum axial ratio at which the mesophase is stable. Such a mesophase is 'absolutely stable' with no thermotropicity, i.e. its nematic → isotropic transition temperature

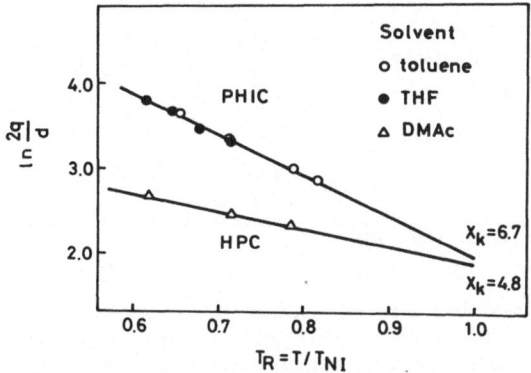

FIG. 2. Temperature dependence of the axial ratio $x = 2q/d$ for poly($n$-hexyl isocyanate) (PHIC) and hydroxypropyl cellulose (HPC) in the indicated solvents. From Reference 6.

$T_{NI}^{\circ} = \infty$. Low molecular weight thermotropic mesogens have finite $T_{NI}^{\circ}$ because, in spite of insufficient geometrical asymmetry ($x < 6.4$), they exhibit asymmetric attractive interactions.[5] Do thermotropic polymers exist by virtue of a similar mechanism?—not necessarily. Results[6] illustrating the temperature variation of the axial ratio for hydroxypropyl cellulose and poly($n$-hexyl isocyanate) are illustrated in Fig. 2. The axial ratio was determined from the persistence length, as required by the treatment of semi-rigid chains.[7] It is seen that at the $T_{NI}^{\circ}$ temperature all polymers have an extrapolated axial ratio close to the critical value, suggesting that thermotropicity occurs when the axial ratio falls below 6.4. Recently,[8] even rigid polyesters such as those formed from terephthalic acid and phenylhydroquinone were shown to exhibit a measurable $T_{NI}^{\circ}$ ($\sim 475°C$) by fast DSC. From the point of view of fiber spinning, these results suggest that all polymers which are melt-spun should be regarded as semi-rigid. In fact, even if the rigidity of the *para*-linked aromatic polyamides and polyesters is similar at a given temperature, as shown by Hummel and Flory,[9] the former, spun as a lyotropic system at room temperature, will exhibit the largest molecular rigidity.

## 1.2. Order Parameter

For rigid lyotropic polymers, theory predicts an increase of the order parameter with both concentration and flow gradient. Both effects have been calculated for spatially uniform systems (monodomain) and are illustrated in Fig. 3(a)[10] which represents the variation of Flory's disorientation parameter, $Y$, for rods with axial ratio 100.[10] Note that the

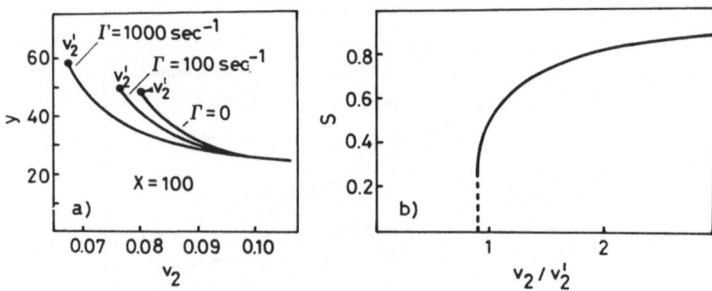

FIG. 3. (a) Flory's disorientation parameter vs. polymer volume fraction for different values of the parallel velocity gradient; the critical $v_2'$ are indicated. Data from Reference 10. (b) Order parameter vs. reduced polymer volume fraction. From Reference 13.

elongational gradient decreases the critical concentration and is effective in increasing order mostly in the more dilute solutions. At large $v_2$, the largest contribution to orientation is that due to the increase of polymer concentration. The latter has been confirmed by recent experimental results.[11] Similar effects have been calculated by Khokhlov and Semenov[12] (in elongational flow) and by Doi[13] (in shear flow using the molecular dynamics approach). The variation of order parameter with concentration calculated by Doi is illustrated in Fig. 3(b). From the standpoint of fiber spinning, these effects should be important for their role in the bulk viscosity. The translation of Doi's result into a viscosity–concentration relationship is illustrated in Fig. 4(a), and is compared in Fig. 4(b) with experimental data[14] for PBA solutions. The viscosity drops discontinuously for $v_2 > v_2'$ and then gradually decreases, reflecting the increase of $S$ with concentration. Only part of these theoretical expectations is reflected in the experimental behavior. The critical concentration, $C_p'$, and its conjugate, $C_p''$, obtained from plots of the volume fraction of isotropic phase vs. overall $C_p$,[4] do not coincide (as was earlier postulated[15]) with the maximum and minimum in the viscosity curve. The increase of $\eta$ past $C_p'$ has been attributed[16,17] to a reinforcement of the isotropic phase by droplets of the anisotropic one, but there is no good theory to explain the increase of $\eta$ at high $C_p$. Crowding effects,[16] 'log jam' effects,[18] and gelation (at $C_p$ slightly greater than that of the minimum, the unsheared solution, cf. Fig. 4(b), began to crystallize) are involved. Note that very high shear rates (close to those occurring at the spinnerette) considerably lower the viscosity. The decrease of $\eta$ with $C_p$ following the transition eventually disappears (an

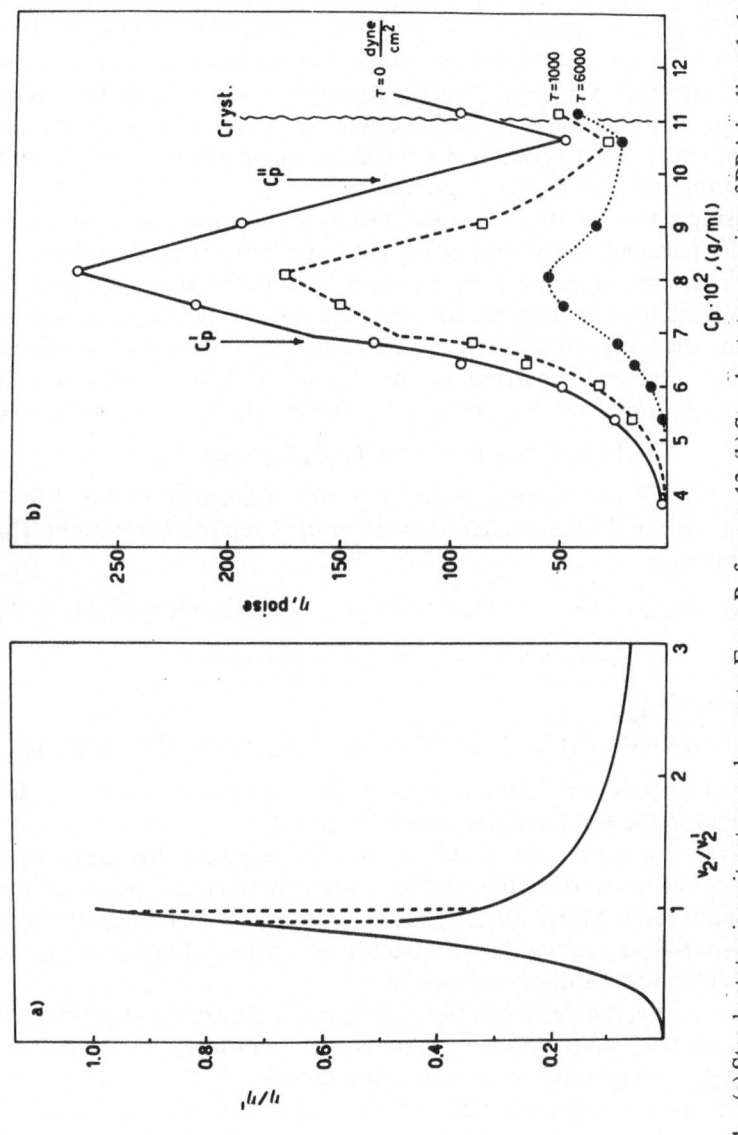

FIG. 4.   (a) Steady state viscosity at zero shear rate. From Reference 13. (b) Steady state viscosity of PBA in dimethylacetamide + 3% LiCl vs. polymer concentration, $C_p$, at fixed shear stress, $\tau$ (partial data from Reference 14; $C_p'$ and $C_p''$ as determined in Reference 4); the wavy line indicates the concentration at which crystallization occurs in the absence of flow.[14]

effect well shown by the data of Kiss and Porter[19] for poly($\gamma$-benzyl-L-glutamate). Under these conditions, $C_p'$ and $\eta$ may be so much lowered that the transition goes undetected, and $\eta$ regularly increases with $C_p$, revealing the persistence of crowding effects. For rigid polymers, log jam effects[18] (even in the nematic phase) could lead to the expectation of an extremely high melt viscosity, possibly leading to cessation of flow. The observation that undiluted rigid polyesters have instead low viscosities (see below) could be justified, as discussed above, in terms of a less rigid conformation at the processing temperature.

Decreasing chain rigidity generally results in a considerable decrease of the order parameter. For instance, Ronca and Yoon[20] predicted values of $S$ at the transition temperature between 0·4 and 0·7 which are greater than the values usually reported for low molecular weight liquid crystals, but smaller than expected for rigid rods. Khokhlov et al.[21] also arrived at similar results and predicted smaller values of $S$ for the worm-like than for the Kuhn chain. Experimentally, for the thermotropic polyester

$$\pm(CH_2)_5-COO-C_6H_4-COO-C_6H_4-OCO\}_x$$

($C_6H_4 = p$-phenylene), Liebert et al.[22] reported $S$ between 0·64 and 0·54 decreasing with molecular weight. Sigaud et al.[23] reported $S = 0·6$ at $T_{NI}^\circ$ for the polyester

$$\pm(CH_2)_{10}-O-C_6H_4-COO-C_6H_4-O-(CH_2)_{10}-O-C_6H_4-OOC-C_6H_4-O\}_x$$

while Noel et al.[24] reported $S \sim 0·3$ for the copolymer

$$\pm OC-C_6H_4-O\}_{0·6}$$
$$\pm OOC-C_6H_4-OCH_2-CH_2-O-C_6H_4-COO-CH_2-CH_2\}_{0·4}.$$

For segmented polymers it might be necessary to consider different order parameters for the rigid and the flexible segment.

These reductions of the order parameter suggest that semi-rigid polymers may be more difficult to orient than rigid ones at the monodomain level. Moreover, a reduced order parameter suggests more frequent spatial variations of the director, or smaller domain sizes, the effect of which will be discussed below.

Experimentally, the determination of the order parameter is performed by X-ray, NMR, anisotropy of magnetic susceptibility, polarized IR spectroscopy using uniform mono-liquid crystals.[11] The order of the mesophase is generally preserved during crystallization. In some cases, solid fibers exhibit a degree of order unexpected from the nematic precursor. Such is the case of the radial crystallite orientation (LCO) characterized, for instance, by electron diffraction or by the refractive index anisotropy on PPTA fiber sections.[25] The latter reveals a radial orientation of crystallite 020 planes while one might have expected the conventional random orientation of crystallites with respect to the radius. LCO could be due to the occurrence of a larger degree of order than uniaxial nematic in the mesophase. Alternatively, LCO may be due to the mode of crystallization in the flow field, and not necessarily be related to the mesophase. More investigation is needed.

### 1.3. Texture

The data in Fig. 4(b) indicate strong non-Newtonian behavior at high shear for the anisotropic solutions. Asada et al.,[26] investigating the rheoptical properties of rigid polypeptides at low shear rates ($\dot{\gamma}$ up to $\sim 100\,s^{-1}$), emphasize the occurrence of the two additional regions indicated in Fig. 5 (region I was also detected by Papkov et al.[27] using PBA solutions, when $\tau$ was below $\sim 10\,dyn\,cm^{-2}$). Asada et al. suggested that a change from a polydomain to a monodomain texture occurs as a result of shear. At low $\dot{\gamma}$ there is domain rotation and slippage followed, at higher $\dot{\gamma}$, by domain deformation and, at still higher $\dot{\gamma}$, by formation of a homogeneous mono-liquid crystal. Thus, region III should be the region where the considerations of the preceding section, and Doi's theory, are more likely to apply. The stability of the polydomain texture under flow was found to be larger the greater the sample thickness and the polymer concentration. Upon cessation of flow, the polydomain texture was generally recovered. A faster relaxation was observed with increasing shear rate. Thermotropic polyesters[28] also showed an evolution of their quiescent texture to an 'ordered texture', and a recovery of the former after shear cessation.

   At intermediate values of shear rates several investigators[19,28-34] have

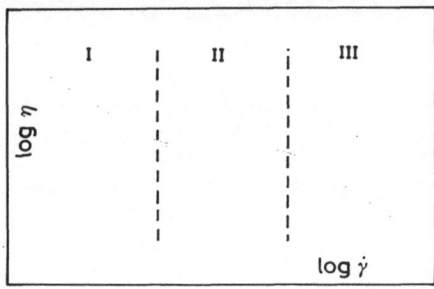

FIG. 5.    Shear viscosity of anisotropic solutions according to Onogi and Asada.[26]

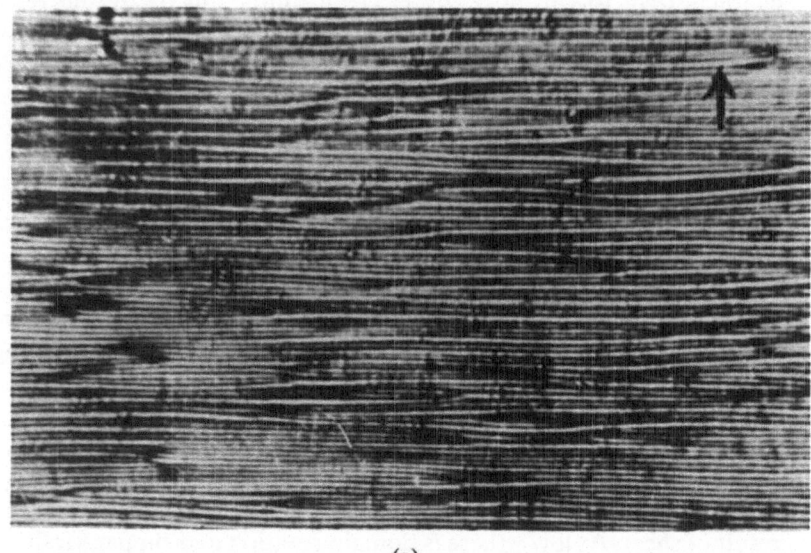

(a)

(b)

FIG. 6. Banded texture in (a) sheared PPTA solution and (b) PPTA fiber. (c) Spherulite grown from unsheared PBA solution (obtained by Chen Shouxi, courtesy of Prof. Qian Renyuan, Academia Sinica, Peking). (d) Schematic representation of lateral crystallite orientation on a section of PPTA fiber 020 plane direction.[25]

(c)

(d)

FIG. 6—contd.

observed the formation of a texture characterized by bands aligned perpendicular to the flow direction (Fig. 6(a)). The texture appears to develop either as a transient during flow field attenuation, or by increasing the shear rate. This texture has been observed with both lyotropic and thermotropic systems, and could be frozen into fibers of PPTA (Fig. 6(b)) and several other lyotropic polymers,[33b] as well as into polyester films. The bands reflect regular (zig-zag) changes in the alignment of the molecules with the flow direction. In fact, by rotation of the sample under the polarizing microscope, dark and bright band extinction occurs[30] as reported by Chinese workers, among the first to observe the banded texture. Zachariades et al.[34] have suggested an interesting correlation between the domain and the banded textures. They start with the observation that the size of the domain at rest $(5-10\,\mu m)$[35] is comparable with the band spacing. Since the relaxation of the domain[26] and of the banded texture[19] is faster the larger the shear rate, they argue that a deformation into smaller domains occurs already at intermediate $\dot{\gamma}$ values. A similar conclusion was reached by Marrucci[36a] who has attempted a description of the band texture (and associated negative normal stresses) in terms of the Leslie–Ericksen theory for the dynamics of conventional nematogens, accounting for the viscoelastic characteristics of polymers. A relationship between domain size, $a$, and domain relaxation time, $\lambda_F$, may be written:[36]

$$\lambda_F = \bar{\eta}a^2/\bar{K} \qquad (2)$$

where $\bar{\eta}$ and $\bar{K}$ are the averages of the six viscosity coefficients and of the three elastic constants of the liquid crystal (the relaxation time of the banded texture was greater than $10^3$ s). Both Marrucci and Zachariades et al. suggest that the smaller domains become oriented to yield the banded texture. The band density increases with shear rate,[34] with spacings decreasing from $1-10\,\mu m$ at $\dot{\gamma} \approx 100\,s^{-1}$ to $0\cdot3-0\cdot8\,\mu m$ at $\langle\dot{\gamma}\rangle \approx 10^3\,s^{-1}$. The latter is comparable with the rate prevailing during fiber extrusion, and therefore domain and band texture appear to be relevant to the fabrication processes of liquid crystalline polymers. The band spacing in PPTA fibers was $\sim0\cdot5\,\mu m$.[29] It should be noted that the formation of small domains under shear appears in contrast with the formation of monodomains at large $\dot{\gamma}$ postulated by Asada et al. It is possible that the formation of monodomains (and hence the highest modulus) is facilitated by elongational, rather than shear, flow.[36,37] On the other hand, experimental details pertaining to the formation of the band texture, particularly the role of increasing or attenuating the flow field, need clarification. In fact, Horio[33b] suggested that formation of the banded texture in fibers is simply due to attenuation of the flow field along the spinning line.

Zachariades et al. noted that thick polyester samples $(>200\,\mu m)$ did not exhibit the band texture and had poor mechanical properties. It is important to point out that a decrease of mechanical properties with increasing sample thickness,[38] and skin and core morphologies,[37] have been observed. The zig-zag angle of molecular axes to the flow direction may tend to negligible values in the skin of well oriented fibers, but still be appreciable ($\sim 10°$)[33b] in the fiber core. The oriented skin may therefore be closer to the monodomain texture than the relatively less oriented core. The problem of achieving maximum orientation and mechanical properties thus requires a detailed knowledge of the nature of domains and of their interaction with the shear and elongational field.

Domain size should also be sensitive to molecular parameters, i.e. the degree of rigidity as discussed above. Thus, in view of eqn (2), semi-rigid polymers could be more difficult to orient than rigid ones, due not only to their smaller order parameter but also to a smaller domain size.

In spite of the growing interest in assessing the role of texture on mechanical properties, quantitative studies of domain size and stability, and the role of molecular parameters, have been scarce. Unfortunately, the study of domain size is not a simple one. Disclinations are difficult to control and to reproduce. The orienting effect of the surface is one of several variables which control texture. The domain size in bulk would require the study of thin slices cut from suitable samples. Such studies are at present in progress.

Phenomenologically, disclinations such as those of strength $\pm 1/2$ typical of nematics, are related to the basic distortions of splay, twist and bend.[1] This type of correlation was qualitatively attempted by Kleman *et al.*,[39] but only in the case of segmented polyesters. Marrucci[36b] attempted to explain the persistence and the recovery of the polydomain texture in terms of a network of defects assumed to preserve its integrity in spite of an orientation and deformation of disclinations under flow. His calculations reproduce semi-quantitatively the occurrence of regions I and II (note that alternative interpretations of region I involve aggregation and the occurrence of yield stress[27]). Marrucci speculates that small defects remain 'hidden' even after extensive domain deformation, and could provide a recovery of the polydomain texture even for samples subjected to the large shear rates involved in processing.

## 2.  SPINNING RESULTS

### 2.1. Rigid Lyotropic

*2.1.1. As-spun Fibers*
Figure 6 includes a photomicrograph indicating that, out of an unsheared PBA solution exhibiting the schlieren pattern, spherulites may develop

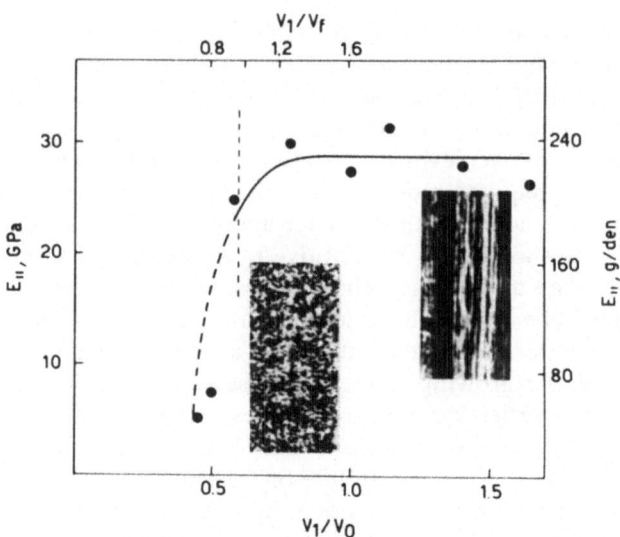

FIG. 7. Initial modulus and optical micrographs of PBA as-spun fibers as a function of pull-off ratio $V_1/V_0$; $V_0$ (outflow rate) $= 10 \, \text{m min}^{-1}$; polymer concentration of the anisotropic dope $= 9 \, \text{g}/100 \, \text{ml}$; $V_f$ is the velocity of the freely extruded filament. From Reference 40. $1 \, \text{GPa} = (100/9\rho) \, \text{g den}^{-1}$.

upon cooling. These negative spherulites are composed of a crystallo-solvate with molecules tangentially oriented.[30] Disoriented crystalline morphologies are also evident in Fig. 7 which reveals the effect of the pull-off ratio on the modulus and on the fibrous pattern of PBA.[40] Anisotropic dopes ($C_p = 9\%$) in dimethylacetamide (DMAc) $+ 3\%$ LiCl were spun in $H_2O$ without an air gap, at an outflow rate $V_0$ of $10 \, \text{m min}^{-1}$, and collected after the coagulation and washing baths at a take-up rate $V_1$. The pull-off, or spin-stretch ratio, is usually defined by $V_1/V_0$. However, due to a small amount of die swelling, the velocity of the freely extruded filament, $V_f$ (no take-up), is somewhat smaller than $V_0$ and therefore $V_1/V_f$ gives a better indication of the tension along the spin-line. We see (Fig. 7) that at very low pull-off (the fiber barely under tension) the modulus is very low, and the fibrous pattern does not develop. Since anisotropic dopes of rigid molecules are characterized by a large microscopic order parameter,[11] we conclude that the domains, not the molecules, need to be aligned before orientation and modulus may increase. Note the sudden increase of modulus from $\sim 5 \, \text{GPa}$ to $\sim 30 \, \text{GPa}$ when $V_1/V_f$ changes a little around unity. This indicates a very unusual and extremely pronounced tendency toward orientation under minimum pull-off. In fact, we regard

Fig. 7 as the best evidence of the importance of a nematic dope in obtaining ultra-high modulus. For no other polymer solution would a small change in the $V_1/V_f$ ratio produce such dramatic effects on the modulus. Under the conditions in Fig. 7, further increase of pull-off ratio has a negligible effect, probably due to the incidence of crystallization (see below). We can now fix the pull-off ratio, say at 1, and explore the effect of dope concentration as shown in Fig. 8. New data for PBA[41] and for PPTA as-spun fibers are compared. (An air gap was not used; at the relatively low molecular weight used for PBA,[41] use of an air gap would have resulted in droplet formation.) The latter data are those of Weyland.[42] The dashed lines in Fig. 8 represent the boundary of the biphasic gap. For PBA, the $C'_p$ boundary was assessed as discussed in connection with Fig. 4(b). For this polymer we observe a sudden increase of the modulus upon crossing the boundary between isotropic and anisotropic phases, but not much effect at still larger $C_p$. For PPTA we also note that the modulus does not increase with concentration at high $C_p$. At variance with PBA, no sudden increase of the modulus is observed upon entering the biphasic region. However, the $C'_p$ boundary was improperly deduced from a plot of $\eta$ vs. $C_p$ at low shear reported in the Weyland paper. Taking $C'_p$ from the shoulder, rather than from the maximum, of the latter curve,[42] we can shift the $C_p$ boundary at lower polymer concentration (as indicated in Fig. 8), and a jump in the modulus then becomes evident. In general, we believe that the location, and even the lack, of a jump at $C'_p$ on the modulus vs. $C_p$ curve should not be regarded as compelling evidence for the role of the mesophase on the high modulus. In fact (cf. Fig. 4), $C'_p$ is determined at low shear, while spinning[40] occurs at $\dot{\gamma}$ values above $10\,000\,\text{s}^{-1}$. In connection with Fig. 8, the significant question to ask is why the modulus tends to level off, or even decrease, when $C_p > \sim 10\%$. Consideration of Fig. 4(b) offers a plausible answer. We see that the useful concentration range for spinning PBA before crystallization occurs is a rather limited one, and is further decreased at high shear rates. The effects responsible for the increase of $\eta$ with $C_p$ past $C'_p$ (discussed in Section 1.2) may also be responsible for the failure to observe the continuous increase of the modulus with $C_p$ which would be expected from the corresponding increase in the order parameter (cf. Fig. 4(a)). Should one then say that the mesophase is of limited or no utility in achieving high modulus, as suggested[42] by Weyland?—certainly not, at least in the case of the rigid lyotropic systems, when $C'_p$ is rather low, and it is not convenient to spin the diluted solutions which occur below $C'_p$. In fact, the packing requirements of an ensemble of rigid rods imply that a solution with $C_p > C'_p$

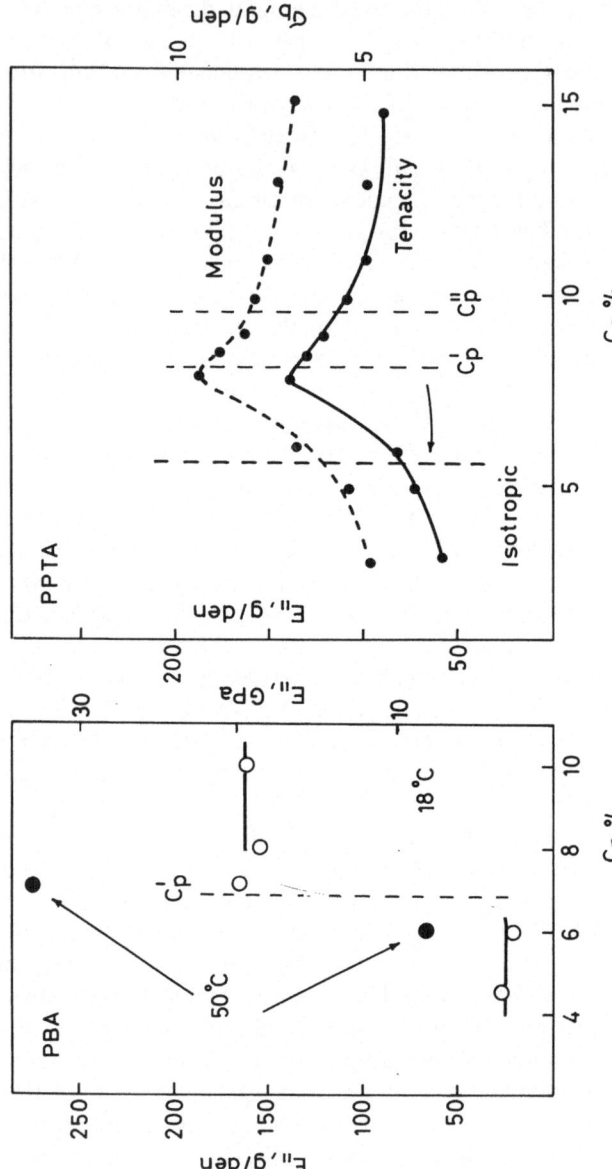

FIG. 8. Initial modulus and tensile strength, $\sigma_b$, for as-spun PBA monofilaments (left, solvent DMAc + 3% LiCl, $V_1/V_f = 1$ die = 100 $\mu$m; data from Reference 41) and for PPTA yarns (right, $\eta_{in} = 4.2$, 99.8% $H_2SO_4$, 40°C, $V_1/V_0 = 1$; from Reference 42) vs. dope concentration ($V_0$ constant = 8 m min$^{-1}$ for PBA, 20 m min$^{-1}$ for PPTA); dashed lines indicate the boundaries of the biphasic region.

must inescapably be an anisotropic one.[3] Moreover, the orientation process of these solutions is based on concepts different from those pertaining to conventional fiber spinning. The use of an air gap, the optimization of $V_0$, $V_1$, the solvent or the temperature of the bath, may be particularly important in situations in which the saturation line is close to the biphasic gap in order to ensure that crystallization sets in when the orientation of the mesophase is already developed. In fact, by using optimal operating conditions, Kwolek[25] reports as-spun moduli up to $\sim 50$ GPa which are larger than those (up to 25 GPa) reported in Fig. 8 without precautions. Data in Fig. 8 show the significant effect of increasing the bath temperature (from 18 to 50°C) on the modulus of PBA.[41] The air gap delays the increase of $C_p$ in the coagulation bath and also assures a large incidence of elongation flow which, as discussed in Section 1, is particularly effective in promoting monodomains. The solvent used is of particular importance since complex formation with the polymer may occur. The solubility of PPTA appears to be enhanced by the formation of crystallosolvates which melt below the pure polymer ($\sim 80$°C when $C_p \sim 20\%$).[25,43a] Crystallosolvates have been reported[43b] even for the PBA/DMAc + LiCl system. The stability of PPTA in 99·8% $H_2SO_4$ appears to be related to the formation of soluble complexes at the spinning temperature (80–100°C). It is noteworthy that one has to reach a 20% solution to maximize stability, fiber tenacity (up to 20 g den$^{-1}$) and radial crystallite orientation (Fig. 6(d)).[25]

It will be shown in the subsequent cases that high modulus can also be achieved outside the region of mesophase stability, i.e. at $C_p < C_p'$ (case B), or by spinning at temperature $T_s < T_{NI}^\circ$ (case D). It is therefore interesting to consider in more detail the results obtained by spinning rigid lyotropic systems below $C_p'$. Figure 8 indicates that in this case moduli of the order of 6 GPa (50 g den$^{-1}$) are obtained at $V_1/V_0 = 1$. The Kwolek patent[25] reports the results reproduced in Table 1 which show a similar value of the

TABLE 1
PROPERTIES OF AS-SPUN PBA FIBERS[a]

| Source | $E_\parallel$ | | $\sigma_b$ | | $\varepsilon_b$ | den |
|---|---|---|---|---|---|---|
| | (g den$^{-1}$) | (GPa) | (g den$^{-1}$) | (GPa) | (%) | |
| Isotropic layer | 64 | 7·7 | 1·2 | 0·14 | 9·0 | 22·5 |
| Anisotropic layer | 283 | 34 | 7·2 | 0·86 | 8·1 | 4·8 |

[a] Data from US Pat. 3,671,542. Re.30,352. Example 65.

TABLE 2

PROPERTIES OF AS-SPUN CELLULOSE FIBERS (REFERENCE 46)[a]

| $C_p$ | $V_0$ | $V_f$ | $V_1$ | $V_1/V_f$ | Diam. ($\mu m$) | $E_{\parallel}$ 0·1% | | $\sigma_b$ | | $\varepsilon_b$ |
|---|---|---|---|---|---|---|---|---|---|---|
| | | ($m\,min^{-1}$) | | | | ($GPa$) | ($g\,den^{-1}$) | ($GPa$) | ($g\,den^{-1}$) | |
| 13·4 | 26 | 7·6 | 10·65 | 1·42 | 49 | 4·3 | 31·4 | 0·20 | 1·46 | 13·3 |
| 13·4 | 26 | 7·6 | 13·05 | 1·74 | 45 | 5·0 | 36·6 | 0·24 | 1·75 | 13·0 |
| 13·4 | 26 | 7·6 | 14·05 | 1·93 | 44 | 5·7 | 41·7 | 0·18 | 1·32 | 8·4 |
| 13·4 | 26 | 7·6 | 17·4 | 2·26 | 38 | 7·4 | 54·1 | 0·29 | 2·12 | 9·7 |
| 13·4 | 26 | 7·6 | 19·5 | 2·58 | 38 | 7·4 | 54·1 | 0·25 | 1·83 | 10·0 |
| 9·1 | 26 | 7·6 | 10·65 | 1·42 | 41 | 7·3 | 53·3 | 0·18 | 1·32 | 17·3 |
| 9·1 | 26 | 7·6 | 13·05 | 1·74 | 38 | 7·9 | 57·7 | 0·16 | 1·17 | 17·9 |
| 9·1 | 26 | 7·6 | 14·05 | 1·93 | 37 | 7·9 | 57·7 | 0·19 | 1·39 | 23·4 |
| 9·1 | 26 | 7·6 | 17·4 | 2·26 | — | 8·8 | 64·3 | 0·22 | 1·61 | 20·9 |
| 9·1 | 26 | 7·6 | 19·5 | 2·53 | — | 8·8 | 64·3 | 0·20 | 1·46 | 15·9 |

[a] Sample L regenerated cellulose II; $|\eta|_{CED}^{25°} = 1\cdot62\,dL\,g^{-1}$; DP = 290 (cf. Reference 44). Fibers spun from DMAc + 7·8% LiCl in $H_2O$; die diam. = $100\,\mu m$; $\dot{\gamma}_{spin} = 35\,000\,s^{-1}$; $V_R/V_1 = 1$; $\rho = 1\cdot52\,g\,cm^{-3}$. Fiber washed in $H_2O$ for 48 h; dried under vacuum for 24 h at 45°C; stored in desiccator until tested.

modulus. (Also the strength, $\sigma_b$, and the elongation to break, $\varepsilon_b$, of PBA fibers are rather poor when spinning is performed from the isotropic solution just close to $C_p'$.) Unfortunately, the Du Pont patent is generally not explicit about the value of $V_1/V_0$ for their as-spun fibers. We expect that optimization of spinning parameters might improve the modulus of as-spun fibers prepared from isotropic dopes due to the occurrence of a flow-induced mesophase[10] (cf. Fig. 3(a)). This approach, which is still based on the occurrence of the mesophase, has not been adequately exploited. However, as we shall see, it is of particular interest for semi-rigid lyotropic systems due to their larger $C_p'$. For PBA, results of more recent experiments[41] appear to suggest that no high modulus is obtained from isotropic dopes for a wide range of $V_1$ and $V_0$, but an improvement begins in the pre-transitional range.

### 2.1.2. Post-spinning Treatments

Thermal treatments[25] have been reported to increase the modulus of fibers as-spun from anisotropic dopes up to $\sim 120\,\text{GPa}$. The effect of these treatments is therefore a relatively modest one with an improvement by a factor of the order of 2 or 3. This reflects the fact that the main orientation was achieved through the mesophase. Thermal treatments on fibers as-spun from isotropic solutions produce, on the other hand, much more striking effects. The Kwolek patent reports the case (example 43) of a fiber prepared from an isotropic dope which exhibited an as-spun modulus of $77\,\text{g}\,\text{den}^{-1}$ ($\sim 9.5\,\text{GPa}$). Upon thermal treatment the modulus reached $800\,\text{g}\,\text{den}^{-1}$. The orientation in this case is primarily due to deformation of a crystalline system composed of rod-like molecules. Measurement of lateral order by the anisotropy of birefringence[25] indicates reduced order when spinning is *not* performed from the anisotropic dope.

### 2.2. Semi-rigid Lyotropic

We consider two polymers, cellulose and the polyterephthalamide of $p$-aminobenzhydrazide (X-500), which show considerable difference in their rigidity, having persistence lengths of $\sim 125$[44] and $50\,\text{Å}$,[45] respectively. The 'pseudo' phase diagram of cellulose in DMAc + LiCl ($C_s$) is reported in Fig. 9. The pure mesophase is reached at rather high $C_p$ and $C_s$ (15 and 8%, respectively), and the line describing the crystallization equilibrium crosses the biphasic region with the result that the useful range for spinning from the mesophase is much smaller than in the case of PBA. Actually, the mesophase occurs in a state of non-equilibrium over most of the composition range. Therefore the benefit of the mesophase cannot be

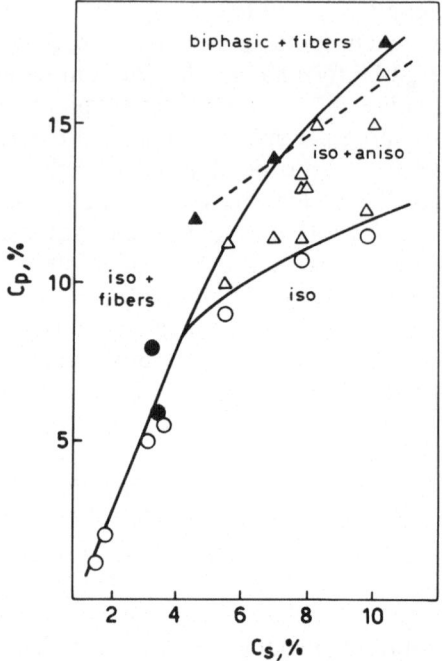

Fig. 9.   Pseudo phase diagram of cellulose (D.P. 288) in DMAc containing varying amounts of LiCl ($C_s$); the dashed line represents the upper limit of the biphasic region ($C_p''$). From Reference 44.

exploited easily for this system, as the results in Table 2 indicate.[46] The modulus obtained by spinning below $C_p'$ is rather low, since spinning conditions are not optimized, but is actually larger than that obtained at the concentration of the mesophase. However, preliminary data show significant improvement by reducing $\dot{\gamma}_{spin}$. For this type of system, the best approach would be that of spinning below $C_p'$, relying on the possibility that the flow-induced transition to the mesophase will occur before crystallization sets in.

Alternatively, solvent systems in which still stronger binding occur may provide the best approach for depressing crystallization and spinning from a mesophase. In $N$-methylmorpholine oxide/water solution, cellulose will form a mesophase[47] from which highly oriented fibers can be produced.[48]

Cellulose derivatives have also been shown to be suitable for the production of high tenacity fibers. In general, however, cellulose

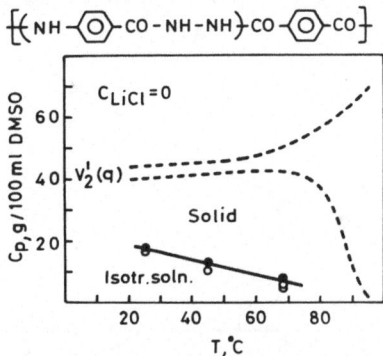

FIG. 10. Phase diagram for X-500 in DMSO; the dashed line indicates that the expected biphasic region is above the solubility limit. From Reference 41.

derivatives attain the mesophase at much higher $C'_p$ than cellulose, due to their larger degree of flexibility. At this high concentration ($> 30\%$) the viscosity is so large that spinning is often problematic. Use of solvents in which $C'_p$ is the smallest possible, or the exploitation of a flow-induced transition, is advisable in such cases.

The case of X-500 in dimethyl sulfoxide (DMSO) (Fig. 10) shows that, on the basis of the persistence length, the mesophase could occur at $C'_p > \sim 40\%$, but it is completely eclipsed by early occurrence of crystallization. Spinning can only be performed from the isotropic solution below $C_p \sim 20\%$, and a lot of effort has been spent to establish whether or not the flow-induced transition occurs.[49-51] The results,[50] summarized in Fig. 11 (no air gap), reveal the following:

(1)  The modulus increases continuously with $C_p$ at given $V_1/V_f$ and may attain $180\,\mathrm{g\,den^{-1}}$ ($\sim 22.5\,\mathrm{GPa}$). Thus, at the as-spun level, the modulus of X-500 may attain the ultra-high level, but it is not as large as that of PBA. Moreover, it increases continuously with $C_p$, with no jump and levelling off.

(2)  The modulus increases with $V_1/V_f$ more gradually than for PBA, at least in the range $1 < V_1/V_f < 1.5$. Thereafter a saturation occurs, just as for PBA. There is one additional difference with respect to PBA: post-spinning treatments are generally more effective (typically, the as-spun modulus of X-500 can be increased by a factor of $\sim 5$ by thermomechanical treatments).

How can we explain these differences? The rheological behavior is illustrated in Fig. 12. The $\eta$ vs. $C_p$ curve exhibits a maximum and a

A. CIFERRI

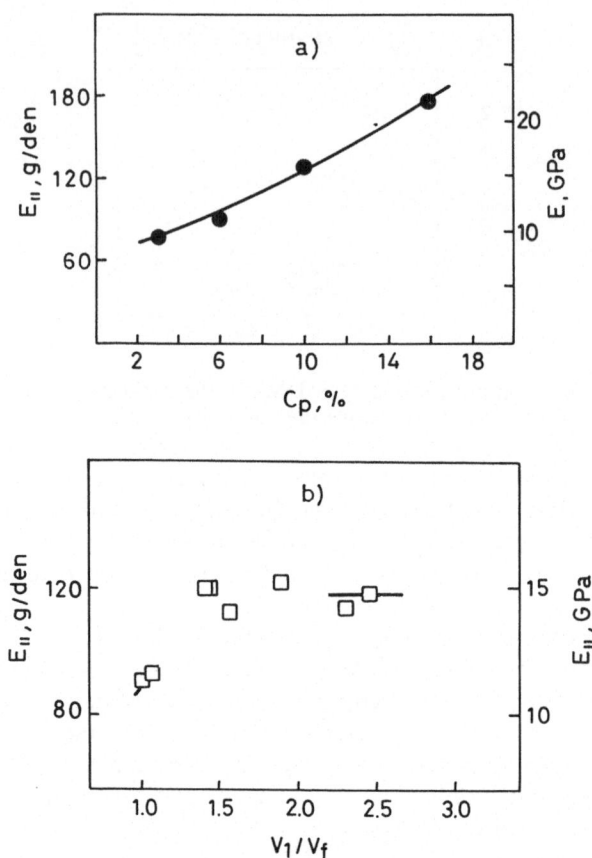

FIG. 11. (a) Initial modulus vs. dope concentration for X-500 fibers as-spun from DMSO; $V_1/V_f = 1.5$. From Reference 41. (b) Effect of pull-off ratio $V_1/V_f$ on the modulus of X-500 fibers as-spun from DMSO ($C_p = 10\%$); only $V_1$ was varied while $V_0$ was kept constant at 11.6 m min$^{-1}$; $V_f$ ($\sim 0.6 V_0$) is the freely extruded velocity of the filament. Data from Reference 50.

minimum which superficially resemble those observed with PBA (Fig. 4(b)), and could therefore suggest the occurrence of a shear-induced mesophase. However, close inspection reveals a marked difference in the effect of the shear rate in the viscosity–concentration plots of X-500 and PBA. A flow instability is probably involved. The log $\eta$ vs. $C_p$ plot at low shear reveals two well defined changes of slope. We attribute the first change, at $C_p^{ent}$, to entanglement formation, as usually observed with flexible polymers. The second change is more speculative and could reflect

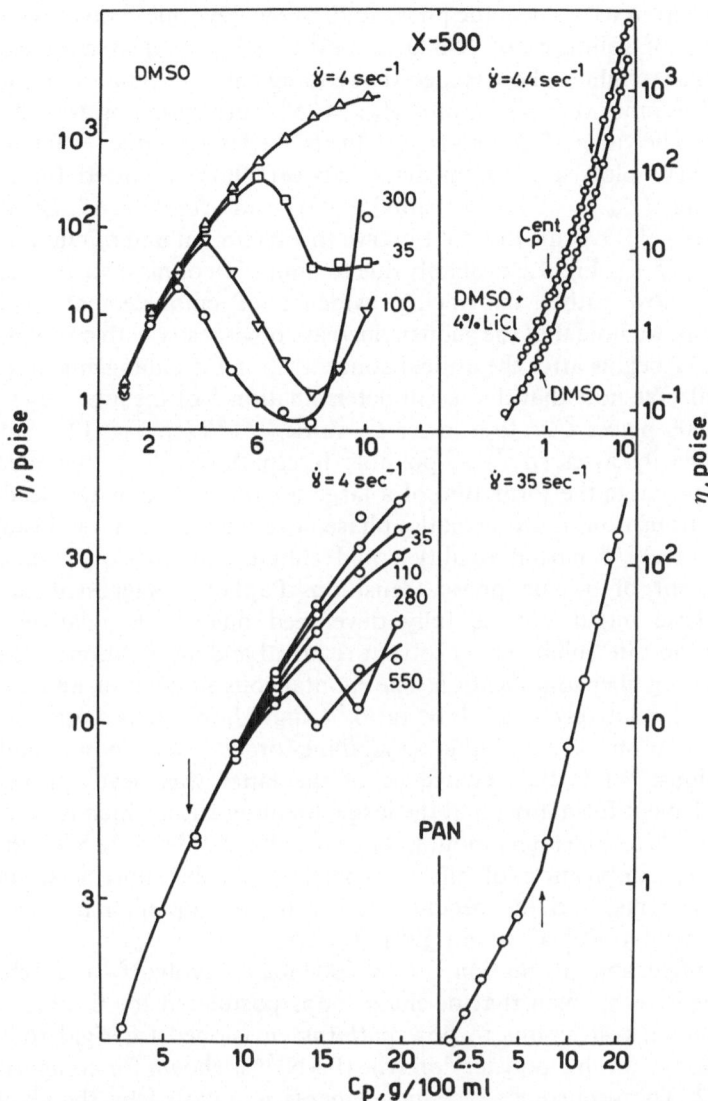

FIG. 12.    Steady-state shear viscosity vs. polymer concentration for X-500 and for PAN at 20°C at the indicated values of shear rate; data at low shear are plotted in log-log plots on the right; solvents are DMSO and DMSO + 4% LiCl for X-500, and dimethylformamide for PAN. All data from Reference 53, except log-log plot for X-500 from Reference 50.

the transition to the nematic phase under flow. Dvornic[52] investigated the viscosity of solutions of X-500 in DMAc ($C_p = 7\%$) as a function of molecular weight and concluded that entanglement formation occurs at a critical number of chain atoms $Z_{WC} = 253$, corresponding to a $M_{WC}$ of $\sim 7000$. The value of $Z_{WC}$ and the value of the largest slope ($\sim 4$) in the log $\eta$–log MW plots are intermediate between those reported for flexible polymers ($Z_{WC} = 350$–650, slope = 3·4) and rigid rods ($Z_{WC} \approx 35$, slope $\approx 7$).[52] (Dvornic did not observe the maximum and minimum in the plot of $\eta$ vs. $C_p$, Fig. 12, probably due to limitation of his instrumentation.) Having now rather convincing evidence of entanglement in X-500 solutions, we note that the sudden decrease of viscosity with concentration in Fig. 12 begins after the critical concentration for entanglement, $C_p^{ent}$, is reached. (We note that the recent determination[50] of the log $\eta$–log $C_p$ plot for X-500 places $C_p^{ent}$ to a lower $C_p$ value, $\sim 1\%$ (cf. Fig. 12), than that preliminarily assessed.[53]) A possible interpretation[50] is that entanglements promote the formation of a large orienting effect which leads to a phase transition to a nematic phase, evolving in a flow instability. Alternatively,[50] partial rigidity could enhance orientation without the attainment of a true phase transition. Papkov[49] suggested that the mesophase might not be fully developed due to its relatively slow nucleation rate enhanced by solvent removal and attenuation of the flow field during fiber solidification. The spontaneous elongation, and the large effect of a relatively small draw ratio,[50] might however be indicative[49] of the formation of the liquid crystalline order under non-equilibrium conditions. While the occurrence of the latter was never proven, the entanglement formation, and the large orienting effect which is eventually responsible for the high modulus, appear to be real. Thus, X-500 exhibits a clear superimposition of effects typical of flexible polymers, such as entanglements, and the peculiar ability to develop high modulus and orientation typical of more rigid polymers.

Entanglement formation in X-500 may involve[52] a mechanism intermediate between that of chain loops postulated for flexible chains, and mutual restrictions to flow patterns envisioned for rigid rods. The behavior of flexible polyacrylonitrile (PAN)[53] is shown for comparison in Fig. 12. The occurrence of entanglements is revealed by the change of slope in the log $\eta$ vs. $C_p$ plot, but no maximum and minimum are apparent. PAN is spun industrially at $C_p \sim 15\%$. At this concentration so many entanglements of the chain loop type are present as to make unlikely the development of the well oriented and organized structure which appears to occur with X-500.[54] It is thus possible that the non-loopy nature of entanglements occurring in semi-rigid polymers facilitates the attainment of high orientation, even in the absence of a transition. In fact, partly 'disentangled' systems based on conventional polymers can develop high modulus and orientation.[55] We are here referring to the 'gel spinning' process of polyethylene, described in another chapter of this book.

## 2.3. Rigid Thermotropic

Poly(p-phenylene terephthalate), with a persistence length of 785 Å, even greater than that of PBA (410 Å),[9] cannot be spun due to its high melting temperature. Asymmetrical substitution reduces $T_{KN}$ and may also affect rigidity, although, as we have seen, it is the use of high temperature which primarily reduces rigidity. The phenyl substituted polymer PPHQT has been reported[56,57] to develop ultra-high modulus,

PPHQT

up to 50 GPa as-spun, increasing up to 105 GPa upon post-polymerization (Table 3). A value of ~ 10 for the $V_1/V_0$ ratio was reported.[56] This is very small compared to the ratio for flexible or semi-rigid (see below) polyesters, reflecting the ease of orientation of the mesophase. Moreover, for practical reasons, both $V_0$ and $V_1$ must be high ($V_1 \rightarrow 1000\,\text{m min}^{-1}$).

We have investigated in more detail the conditions for spinnability of

TABLE 3

FIBER PROPERTIES OF PHENYLHYDROQUINONE HOMOPOLYESTERS (PATENT AND LITERATURE DATA)

| | As-spun | | Heat-treated 60 min at 340°C, Ref. 57 |
|---|---|---|---|
| | Ref. 56[a] | Ref. 57[b] | |
| $\sigma_b$, g den$^{-1}$ (GPa) | 3·4 (0·4) | 3·8 (0·45) | 32 |
| $\varepsilon_b$ (%) | 0·9 | 0·9 | 4·3 |
| $E_{\parallel}$, g den$^{-1}$ (GPa) | 408   (47) | 440   (51) | 910 |
| Denier | 400 | 1·5 | 1·4 |
| O.A. | 14° | — | — |

[a] $T_s = 346°C$; $\eta_{inh} = 1·0$.
[b] $T_s = 340°C$; $\eta_{inh} = 3$ ($M_w \approx 8000$) before heat treatment.

TABLE 4
DSC DATA FOR PPHQT (FROM REFERENCE 58)

| $\eta_{inh}$ (dl g$^{-1}$) | $T_g$ (°C) | $T_{KK}$ (°C) | $T_{KN}$ (°C) | $T_{NI}$[a] (°C) | $\Delta H_{tot}$ (kJ mol$^{-1}$) |
|---|---|---|---|---|---|
| 1·9 | 170 | 320 | 345 | 475 | 5·3 |

[a] Heating rate 40° min$^{-1}$.

PPHQT, without attempting a maximization of its properties.[58] The DSC thermogram (Table 4) reveals a crystalline transition at 320°C, and a $T_{KN}$ at 345°C. The melt viscosity in the Newtonian range (Fig. 13) shows a monotonic decrease with temperature, attaining values below 100 stokes in the nematic region, but showing that the polymer will flow even below $T_{KN}$. Conditions for spinnability were investigated by systematically increasing the spinning temperature, $T_s$, and the extrusion rate, $V_0$, starting from the lowest values (for $V_0$ our available range was between 10 and 650 cm min$^{-1}$). The results (Table 5) indicate that at $T_s < T_{KN}$ it was impossible (even at the highest extrusion rate) to obtain good filaments which could be wound on a bobbin. On the other hand, at $T_s > T_{KN}$ very good spinnability was exhibited. The fiber, which could be collected on a bobbin at a take-up rate ($V_1$) of 100 m min$^{-1}$, exhibited a respectable modulus of $\sim 12$ GPa (100 g den$^{-1}$) and good crystalline orientation. A high extrusion rate ($V_0$) had, however, to be applied.

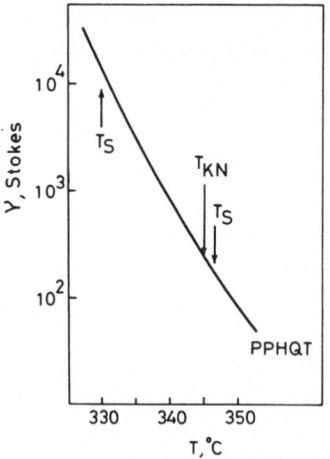

FIG. 13.  Temperature dependence of $v$, the kinematic melt viscosity in the Newtonian range of PPHQT. From Reference 58.

## TABLE 5
SPINNING AND MECHANICAL DATA FOR AS-SPUN PPHQT
(FROM REFERENCE 58)

|  | $T_s = 330°C$ | $T_s = 346°C$ |
| --- | --- | --- |
| $V_0$ (cm min$^{-1}$) | 650 | 650 |
| $V_1$ (m min$^{-1}$) | — | 100 |
| $V_1/V_0$ | — | 15 |
| $E_{\parallel}/\sigma_b/\varepsilon^a$ | 1·2/0·006/1 | 12/0·05/1 |
| Denier | — | 20 |
| O.A. | — | 12° |

$^a$ $E_{\parallel}$, $\sigma_b$ = GPa, $\varepsilon$ = %; die diameter = 750 $\mu$m.

These characteristics of the processability of rigid polyesters appear to be in line with those expected for a nematic system. High modulus and orientation are easily obtained at $T_s > T_{KN}$, and the ratio $V_1/V_0$ need not be large to obtain these advantages. Still, two aspects of these results are not immediately obvious. One is the observation that $V_0$ needs to be rather high to observe good spinnability. A similar situation applies to the spinning of the semi-rigid polyesters at $T_s > T_{KN}$, to be discussed in Section 2.4. Our interpretation is as follows. The melt viscosity of the nematic phase is too low. The tendency toward formation of a droplet, rather than a continuous filament, has to be counteracted by the application of a large flow gradient $\dot{\gamma}$. Even though this may result in a still lower viscosity in the non-Newtonian range, the shear stress ($\eta\dot{\gamma}$) will be large enough to counteract surface tension effects. Instability of the stream due to surface tension ($\pi$) is controlled, according to Ziabicki,[59] by the ratio $\pi/\eta V_0$. From a practical standpoint, the need to apply large $V_0$ to the low viscosity nematic polymers is an advantage. Since these polymers do not allow the application of a large $V_1/V_0$ ratio, a large $V_0$ allows a rather large rate of fiber production.

The second surprising consideration is that a large as-spun modulus was obtained in spite of the expectation (discussed in Section 1) that the axial ratio of PPHQT at $T_s$ should be considerably smaller than that prevailing at room temperature. In fact, $x$ could reach the critical value of 6·4 at 475°C, which is the temperature at which the isotropic state is attained (Table 4). Since, at 25°C, $x$ is of the order of 100, at $T_s \approx 346°C$ $x$ is estimated to be $\sim 15$. Reduced rigidity implies reduced order parameter and perhaps smaller domain size, as discussed in Section 1 (Flory's and Doi's theories predict $S > 0.9$ for undiluted rods, whereas the largest value of $S$

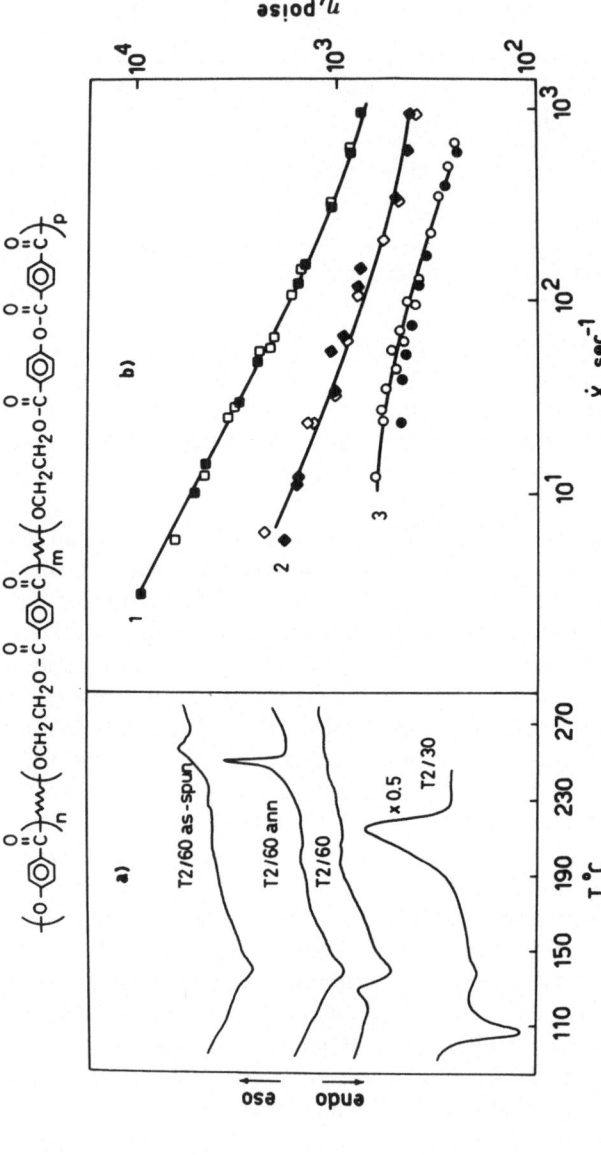

Fig. 14. (a) DSC thermograms for $T_2/60$ and $T_2/30$. From Reference 62. (b) Viscosity vs. shear rate for $T_2/60$ at 210°C: 1, virgin sample; 2, sample exposed for a few minutes at 240°C; 3, sample exposed at 300°C. From Reference 63.

predicted for semi-rigid chains at $T_{NI}^{\circ}$ is $\sim 0.7$). One must conclude that for these polyesters rigidity and domain size are still relatively large at $T_s$. Therefore, relatively long relaxation times, and a rate of crystallization which is probably larger than for lyotropic systems, allow the solidification of an oriented structure. The situation changes drastically as we move to the still less rigid, segmented polyesters.

## 2.4. Semi-rigid Thermotropic

Transesterification of poly(ethylene terephthalate) (PET) with $p$-acetoxybenzoic acid (HBA) yields random copolymers which are nematic when at least 40% HBA residues are incorporated.[38,60,61] DSC data are reported in Fig. 14(a) for copolymers, containing 30 (T$_2$/30) and 60 (T$_2$/60) mole % of $p$-oxybenzoate.[62] T$_2$/30 is not nematic and only shows the melting of PET sequences. T$_2$/60, on the other hand, shows no PET melting but develops upon annealing an endotherm at $\sim 230$–$250°$C which was attributed to the fusion of short HBA sequences.[62,63] The melt viscosity[38,63] of T$_2$/60 in Fig. 14(b) reveals values of $\eta$ of the order of 1000 poise at $\dot{\gamma} \approx 100\,s^{-1}$, which is the largest value observed with our nematic polyesters. Figure 14(b) also reveals non-Newtonian behavior and thermal history effects; $\eta$ at 210°C is strongly decreased when measured after exposing the sample at 240°C or 300°C, an effect attributed to the melting of short HBA crystallites. Other data for the same polymer showed that electrohydrodynamic instabilities[62,64] did not develop below 265°C. Optical microscopy also revealed that the fluidity and the texture typical of nematic polymers developed only above 265°C upon first heating. Thus, T$_2$/60 is a sample with a very low degree of crystallinity and a slow crystallization rate which is basically nematic over a wide temperature range, although the full character of the mesophase is manifested only above 265°C when short PHA sequences are melted.

We tried to spin this sample above and below 265°C, comparing its behavior to that of non-nematic T$_2$/30. The results (Fig. 15) show features which are rather different from those observed with PPHQT, in particular (1) good spinnability is exhibited already at rather low $V_0$ (10 cm min$^{-1}$), (2) the $V_1/V_0$ ratio can be made extremely large, up to 2500, with a gradual increase of the modulus, (3) spinnability and high modulus are obtained well below $T_{KN}$. A more recent report[62b] relates spinnability below $T_{KN}$ to supercooling resulting from thermal history and ester interchange. However, the latter report[62b] does not support the decrease of modulus with temperature illustrated in Fig. 15. High modulus ($E \to 30\,GPa$) was in fact reported in the 260–290°C range.[62b] We note that our values for

A. CIFERRI

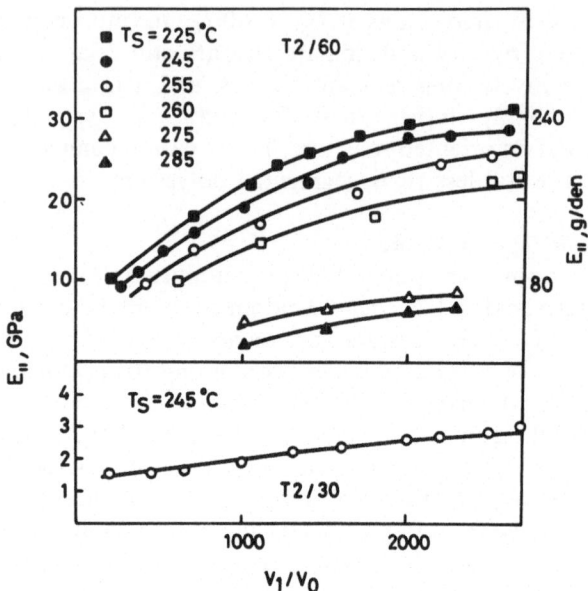

FIG. 15.   Elastic modulus of as-spun $T_2/60$ and $T_2/30$ at the indicated extrusion temperatures vs. the $V_1/V_0$ ratio; $V_0 = 10\,cm\,min^{-1}$; die diameter $= 2\,mm$. From Reference 62.

modulus (and strength)[62] are larger than those reported by Tennessee Eastman investigators[38,60,61] working with thicker samples. They showed that flexural properties improve when sample thickness decreases. Therefore working with thin fibers allows an improvement of orientation and mechanical properties. This may be attributed to the problem of domain orientation in bulk samples discussed in Section 1.

Thus the spinning of nematic $T_2/60$ exhibits several features typical of the more flexible polymers, plus the unique characteristics that ultra-high modulus is obtained when $T_S$ is *below* the temperature at which liquid crystalline behavior is fully apparent. Clearly therefore spinning from a mesophase is not advantageous under all conditions! In general, the freezing of an oriented structure depends upon the occurrence of slow relaxation times in the liquid state, and of fast solidification once orientation is achieved. Jackson[57] measured relaxation times in the liquid crystalline state for a variety of polymers including PET, $T_2/60$ and PPHQT (Fig. 16). He found that relaxation times increased with chain rigidity. The relaxation mechanism of the liquid crystal could be related to the order parameter and to the domain size, both of which are expected to

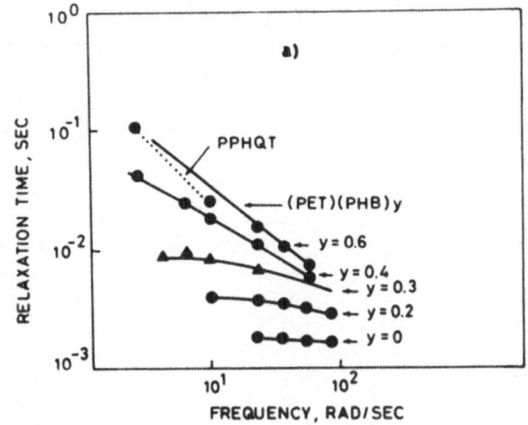

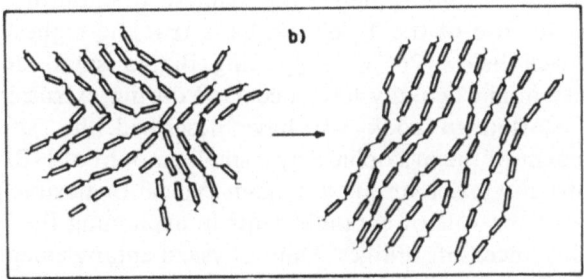

FIG. 16. (a) Relaxation times for polyesters; $T = 360°C$ for PPHQT and 275°C for PET/HBA copolymers. From Reference 57. (b) Orienting semi-rigid chains.

decrease with decreasing rigidity (cf. Section 1), thus demanding a large contribution to orientation from the flow field. As suggested in Fig. 16 (bottom), intramolecular relaxation (e.g. $\lambda \approx 10^{-2}$ s) may also be involved. (A much larger $\lambda_F$, of the order of $10^3$ s, is expected for domain relaxation, complicated also by the effect of shear on domain size; cf. Section 1.3.) There is, in fact, some evidence that semi-rigid molecules do not adopt a rod-like conformation in the nematic phase (i.e. $\Delta S_{KN} > \Delta S_{NI}$).[65] Unfortunately, we do not possess corresponding estimates of crystallization rates for the polymers in Fig. 16, but we know that the degree of HBA crystallinity in $T_2/60$ is extremely low. We can expect that, particularly when crystallization is slow or small in degree, decreasing temperature will increase both relaxation times and crystallization rates, thus favoring the

## TABLE 6
### Properties of injection-moulded 4,4'-(alkylenedioxy)dibenzoic acid polyesters (from Reference 57)

$$\left[\mathrm{C}-\!\!\left\langle\bigcirc\right\rangle\!\!-\mathrm{O(CH_2)}_n\mathrm{O}-\!\!\left\langle\bigcirc\right\rangle\!\!-\mathrm{C}-\mathrm{O}-\!\!\left\langle\bigcirc\right\rangle\!\!-\mathrm{O}\right]_x$$

|                              | $n = 2$ | $n = 3$ | $n = 4$ | $n = 5$ |
|------------------------------|---------|---------|---------|---------|
| Tensile strength (GPa)       | 0·26    | 0·18    | 0·07    | 0·04    |
| Flexural strength (GPa)      | 0·20    | 0·14    | 0·10    | 0·05    |
| Flexural modulus (GPa)       | 11·4    | 6·4     | 8·3     | 2·7     |
| Rockwell hardness (L)        | 102     | 81      | 88      | 34      |
| Heat-deflection temp. (°C)   | 177     | —       | 148     | 74      |
| Polymer melting point (°C)   | 316     | —       | 212     | —       |

preservation of a higher degree of nematic orientation. However, the interesting feature of the $T_2/60$ results is that the highest modulus was obtained well below 265°C, suggesting that a small degree of HBA crystallinity might actually have occurred during extrusion, even before the fiber started to cool. We have described this situation[62] as a 'reinforcement of the mesophase by a small amount of HBA crystallinity'. Such reinforcing mechanisms are again typical of flexible polymers, and remind us of the role of entanglements in explaining the behavior of X-500. For instance, salt bridges[66] and physical entanglements[67] have been advocated for explaining the high orientation of nylon-6–LiCl and polyethylene. Moreover, it is well known that the latter polymer develops high modulus and orientation by solid-state extrusion just below the melting temperature.[68]

As we increase the degree of flexibility of the polyesters through an increase of the number of methylene units, the possibility of developing high modulus tends to disappear. Jackson[57] reported the data in Table 6 for semi-flexible polyesters containing an ether linkage and a variable number of methylene units, $n$. He found that the flexural modulus decreased by a factor of 4 as $n$ was increased from 2 to 5. Note the alternation of $T_{KN}$ and moduli for even and odd members in Table 6.

With specific reference to fibers, we investigated segmented homopolyesters formed from azelaic acid and either 4'-hydroxyphenyl-4-hydroxycinnamate (C-7) or 4,4'-diacetoxybiphenyl (PB-7).[69] DSC data (Table 7) are superimposed on the melt viscosity vs. temperature plots in Fig. 17. The viscosity decreases continuously with temperature for C-7,

## TABLE 7
DATA FOR SEGMENTED HOMOPOLYESTERS PB-7[a] AND C-7[b] (FROM REFERENCE 69)

| Polymer | $\eta_{inh}$ $(dl\ g^{-1})$ | $T_{KN}$ $(°C)$ | $T_{NI}$ $(°C)$ | $T_S$ $(°C)$ | $V_0^c$ $(cm\ min^{-1})$ | $V_1/V_0$ | $E_{\parallel}/\sigma_b$ $(g/GPa)$ |
|---------|------|------|------|------|------|------|------|
| PB-7 | 0·80 | 239 | 250 | 227 | 10 | — | ~0 |
| C-7 | 0·68 | 196 | 285 | 207 | 650 | 40 | 1·7/0·03 |

[a] PB-7 $\text{-[-O-}\langle\bigcirc\rangle\langle\bigcirc\rangle\text{-O-CO-(CH}_2)_7\text{-CO-]-}$

[b] C-7 $\text{-[-O-}\langle\bigcirc\rangle\text{-CO-O-CH=CH-}\langle\bigcirc\rangle\text{-CO-(CH}_2)_7\text{-CO-]-}$

[c] Die diameter = 500 $\mu$m.

but shows a minimum for PB-7 in the transition range, when values of ~100 stokes are attained. PB-7 exhibited some spinnability at $T_S < T_{KN}$ and low $V_0$, but fiber properties were extremely poor (Table 7). C-7 could be spun in the nematic range provided that extremely large extrusion rates were applied ($V_0 = 6\cdot5\,\text{m min}^{-1}$). However, even at $V_1/V_0 = 40$ the modulus remained of the order of $15\,\text{g den}^{-1}$. The need, and the reasons, for the large $V_0$ are similar to those described above for rigid PPHQT. The

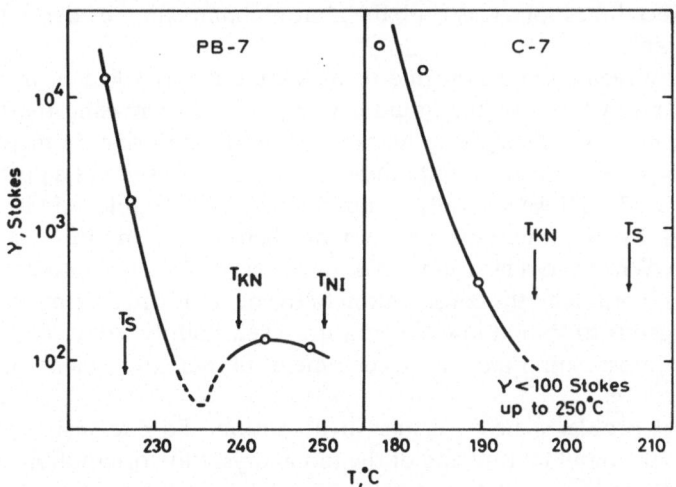

FIG. 17. Temperature dependence of the Newtonian kinematic viscosity of polyesters C-7 and PB-7; transition and spinning temperatures (Table 7) are indicated. From Reference 69.

C-7 polymer, however, shows that satisfaction of the two conditions ($V_0$ large and $T_S > T_{KN}$) does not necessarily produce high modulus. Yet, the inherent and the bulk viscosities of PPHQT and C-7 were similar. We believe that the poor results for the latter polymer are due to too short relaxation times (associated with molecular rearrangement as well as with a small order parameter and domain size). However, even the degree of crystallinity and crystal perfection of semi-flexible nematic polyesters have been found often to be poor, with a molecular conformation not fully extended in the crystalline phase.[65,69]

## 3. CONCLUSIONS

Lyotropic systems may exhibit large molecular rigidity, large $S$ and domain size and, therefore, large relaxation times. However, rigidity also entails a very pronounced tendency to crystallize. The useful composition range for spinning from the mesophase may then be very small, complicating the maximization of orientation for as-spun fibers. Air gaps or 'smart' solvents may be necessary. Decreasing rigidity entails an increase of $v_2'$ and then crystallization encroaches, or even precedes, the mesophase. In these cases, spinning from the isotropic phase and post-spinning treatments still allow the attainment of ultra-high modulus. A flow-induced mesophase is probable, but no compelling evidence for it is yet available.

Rigidity seems important due to its simultaneous effect of increasing the relaxation times of the liquid crystal, and of increasing the melting temperature, thus facilitating quick freezing of the oriented mesophase. Rigid polyesters appear to fulfil these expectations in spite of a decrease in rigidity at $T_s$. However, only a small degree of flexible bonds can be tolerated by the thermotropic systems before the modulus starts to decrease. When the relaxation times, or the crystallization rates, are small and not favorable to the conservation of the nematic orientation, it may be advantageous to spin at low temperature, even slightly below $T_{KN}$, relying on an almost simultaneous development of nematic orientation and crystallization.

Several problems need further investigation. To cite just a few: the origin of the band texture and of the radial crystallite orientation, the role of soluble and crystalline complexes on the rigid lyotropic systems, the role of a higher degree of order than the uniaxial nematic ('cybotactic',[70] or 'biaxial'[71]), the flow-induced transition.

## ACKNOWLEDGEMENTS

The author expresses his appreciation to Prof. K. Katayama and Prof. K. Monobe for the invitation to present the material included in this chapter before the International Symposium on Fiber Science and Technology, and the ensuing Post Symposium, organized by the Society of Fiber Science and Technology, Japan.

## REFERENCES

1. DE GENNES, P. J. (1974). *The physics of liquid crystals*, Clarendon Press, Oxford.
2. ASADA, T. (1982). In: *Polymer liquid crystals* (Ed. A. Ciferri, W. R. Krigbaum and R. B. Meyer), Academic Press, New York.
3. FLORY, P. J. (1956). *Proc. Roy. Soc.*, A234, 73; (1984). *Adv. Polym. Sci.*, 59, 1.
4. CONIO, G., BIANCHI, E., CIFERRI, A. and TEALDI, A. (1981). *Macromolecules*, 14, 1084.
5. FLORY, P. J. and RONCA, G. (1979). *Mol. Cryst. Liq. Cryst.*, 54, 311.
6. KRIGBAUM, W. R., HAKEMI, H., CIFERRI, A. and CONIO, G. (1985). *Macromolecules*, 18, 973.
7. FLORY, P. J. (1978). *Macromolecules*, 11, 1141.
8. KRIGBAUM, W. R., HAKEMI, H. and KOTEK, R. (1985). *Macromolecules*, 18, 965.
9. HUMMEL, J. P. and FLORY, P. J. (1980). *Macromolecules*, 13, 479.
10. MARRUCCI, G. and CIFERRI, A. (1977). *J. Polym. Sci. Polym. Lett. Ed.*, 15, 643.
11. SARTIRANA, M. L., MARSANO, E., BIANCHI, E. and CIFERRI, A. (1986). *Macromolecules*, 19, 1176.
12. KHOKHLOV, A. R. and SEMENOV, A. N. (1982). *Macromolecules*, 15, 1272.
13. DOI, M. (1981). *J. Polym. Sci. Polym. Phys. Ed.*, 19, 229.
14. BALBI, C., BIANCHI, E., CIFERRI, A., TEALDI, A. and KRIGBAUM, W. R. (1980). *J. Polym. Sci. Polym. Phys. Ed.*, 18, 2037.
15. HERMANS, JR, J. (1962). *J. Colloid Sci.*, 17, 638.
16. AHARONI, S. M. (1980). *Polymer*, 21, 1413.
17. MATHESON JR, R. R. (1980). *Macromolecules*, 13, 643.
18. EDWARDS, S. F. and EVANS, K. E. (1982). *J. Chem. Soc. Faraday Trans. 2*, 78, 113.
19. KISS, G. and PORTER, R. S. (1980). *J. Polym. Sci. Polym. Phys. Ed.*, 18, 361; (1980). *Mol. Cryst. Liq. Cryst.*, 60, 267.
20. RONCA, G. and YOON, D. Y. (1982). *J. Chem. Phys.*, 76, 3295.
21. KHOKHLOV, A. R. and SEMENOV, A. N. (1981). *Physica*, A108, 546.
22. LIEBERT, L., STRZELECKI, L. and VAN LEVELUT, D. (1981). *Eur. Polym. J.*, 17, 71.
23. SIGAUD, G., YOON, D. Y. and GRIFFIN, A. C. (1983). *Macromolecules*, 16, 875.
24. NOEL, C., LAUPRETRE, F., FRIEDRICH, C., FAYOLLE, B. and BOSIO, L. (1984). *Polymer*, 5, 808.
25. BLADES, H. (1971). US Patent 2,134,582; (1975). 3,869,429; KWOLEK, S. L. (1980). US Patent 3,671,542 reissued 30,352 (Du Pont Co.).
26. ASADA, T., KODA, T. and ONOGI, S. (1981). *Mol. Cryst. Liq. Cryst.*, 68, 231.

ONOGI, S. and ASADA, T. (1980). In: *Rheology* (Ed. G. Astarita, G. Marrucci and L. Nicolais), Vol. I, Plenum, New York.

27. PAPKOV, S. P., KULICHIKHIN, U. G., KALMYKOVA, V. D. and MALIN, A. Y. (1974). *J. Polym. Sci. Polym. Phys. Ed.*, **12**, 1753.

28. GRAZIANO, D. J. and MACKLEY, M. R. (1984). *Mol. Cryst. Liq. Cryst.*, **106**, 73.

29. SIMMENS, S. C. and HEARLE, J. W. S. (1980). *J. Polym. Sci. Polym. Phys. Ed.*, **18**, 871.

30. SHOUXI, C., CHENFEN, L. and LIYING, C. (1981). *Gaofenzi Tongxun*, **6**, 424.

31. DONALD, A. M., VINEY, C. and WINDLE, A. H. (1983). *Polymer*, **24**, 155.

32. DONALD, A. M. and WINDLE, A. H. (1983). *J. Mater. Sci.*, **18**, 1143.

33. (a) HORIO, M., ISHIKAWA, S. and ODA, K. US–Japan Seminar on Polymer Liquid Crystals, Kyoto, 1983, to appear in *Appl. Polym. Symp.* (b) HORIO, M. (1985). Paper presented at the Post Symposium on Formation, Structure and Properties of High Modulus and High Tenacity Fibers, Kyoto, Japan.

34. ZACHARIADES, A. E., NAVARD, P. and LOGAN, J. A. (1984). *Mol. Cryst. Liq. Cryst.*, **110**, 93.

35. MEETEN, G. H. and NAVARD, P. (1982). *Polymer*, **23**, 1727.

36. MARRUCCI, G.: (a) (1985). *IUPAC Symp. Non-crystalline order in polymers*, Naples, Italy; (b) (1984). In: *9th Int. Congr. Rheology* (Ed. B. Mena, A. Garcia Rejon and C. Range Naisale), Univ. Natl. Aut. de Mexico, Acapulco, p. 441.

37. IDE, Y. and OPHIR, Z. (1983). *Polym. Eng. Sci.*, **23**, 261.

38. JACKSON JR, W. J. and KUHFUSS, H. F. (1976). *J. Polym. Sci. Polym. Chem. Ed.*, **14**, 2043.

39. KLEMAN, M. (1985). *Faraday Discuss. Chem. Soc.*, **79**, 215; KLEMAN, M., LIEBERT, L. and STRZELECKI, L. (1983). *Polymer*, **24**, 295.

40. CIFERRI, A. and VALENTI, B. (1979). In: *Ultra-high modulus polymers* (Ed. A. Ciferri and I. M. Ward), Applied Science Publishers, London; (1978). ALFONSO, G. C., BIANCHI, E., CIFERRI, A., RUSSO, S., SALARIS, F. and VALENTI, B., *J. Polym. Sci. Polym. Symp.*, **65**, 213.

41. CONIO, G., BRUZZONE, R., BIANCHI, E., CIFERRI, A. and TEALDI, A. *Polymer J.* (In press.)

42. WEYLAND, H. A. (1980). *Polym. Bull.*, **3**, 331.

43. (a) GARDNER, K. H., MATHESON, R. R., AVAKIAN, P., CHIA, Y. T. and GIERKE, T. D. (1984). *Polymers for fibers and elastomers*, ACS. (b) TAKASE, M., CIFERRI, A., and KRIGBAUM, W. R. (1986). *J. Polym. Sci. Polym. Phys. Ed.* (In press.)

44. BIANCHI, E., CIFERRI, A., TEALDI, A., COSANI, A. and TERBOJEVICH, M. (1985). *Macromolecules*, **18**, 646.

45. BIANCHI, E., CIFERRI, A., PRESTON, J. and KRIGBAUM, W. R. (1981). *J. Polym. Sci. Polym. Phys. Ed.*, **19**, 863.

46. CONIO, G., *et al.*, unpublished results.

47. CHANZY, H. and PEGUY, A. (1980). *J. Polym. Sci. Polym. Phys. Ed.*, **18**, 1137.

48. FRANK, N. and VARGA, J. (1979). US Patent 4,145,532 (Akzo Co.).

49. PAPKOV, S. P. (1984). *Adv. Polym. Sci.*, **59**, 75.

50. VALENTI, B., ALFONSO, G. C., CIFERRI, A., GIORDANI, P. and MARRUCCI, G. (1981). *J. Appl. Polym. Sci.*, **26**, 3643.

51. BAIRD, D. G., CIFERRI, A., KRIGBAUM, W. R. and SALARIS, F. (1979). *J. Polym. Sci. Polym. Phys. Ed.*, **17**, 1649.

52. DVORNIC, P. R. (1984). *Macromolecules*, **17**, 1348.

53. VALENTI, B. and CIFERRI, A. (1978). *J. Polym. Sci. Polym. Lett. Ed.*, **16**, 657.
54. BLACK, W. B. and PRESTON, J. (1973). *High modulus wholly aromatic fibers*, Marcel Dekker, New York.
55. SMITH, P., LEMSTRA, P. J. and BOOIJ, H. C. (1981). *J. Polym. Sci. Polym. Phys. Ed.*, **19**, 877.
56. SCHAEFGEN, J. R., PLETCHER, T. C. and KLEINSCHUSTER, J. J. (1975). Belg. Patent 829,935 (E. I. Du Pont de Nemours Co.).
57. JACKSON JR, W. J. (1980). *Brit. Polym. J.*, **12**, 154.
58. ACIERNO, D., LA MANTIA, F. P., POLIZZOTTI, G., CIFERRI, A., KRIGBAUM, W. R. and KOTEK, R., unpublished results.
59. ZIABICKI, A. (1976). *Physical foundations of fiber formation*, Wiley, New York.
60. KUHFUSS, H. F. and JACKSON JR, W. J. (1973). US Patent 3,778,410 (Eastman Kodak Co.).
61. KUHFUSS, H. F. and JACKSON JR, W. J. (1975). US Patent 3,804,805 (Eastman Kodak Co.).
62. (a) ACIERNO, D., LA MANTIA, F. P., POLIZZOTTI, G., CIFERRI, A. and VALENTI, B. (1982).*Macromolecules*, **15**, 1455. (b) TEALDI, A., CIFERRI, A. and CONIO, G., *Polym. Commun.* (Submitted for publication.)
63. WISSBRUN, K. F. (1980). *Brit. Polym. J.*, **12**, 163.
64. KRIGBAUM, W. R., LADER, H. J. and CIFERRI, A. (1980). *Macromolecules*, **13**, 554.
65. KRIGBAUM, W. R., WATANABE, J. and ISHIKAWA, T. (1983). *Macromolecules*, **16**, 1271.
66. ACIERNO, D., LA MANTIA, F. P., POLIZZOTTI, G. and CIFERRI, A. (1979). *J. Polym. Sci. Polym. Phys. Ed.*, **17**, 1903.
67. MARRUCCI, G. (1975). *Polym. Eng. Sci.*, **15**, 229.
68. KOGIMA, S. and PORTER, R. S. (1978). *J. Appl. Polym. Sci. Appl. Polym. Symp.*, **33**, 129.
69. ACIERNO, D., LA MANTIA, F. P., POLIZZOTTI, G., CIFERRI, A., KRIGBAUM, W. R. and KOTEK, R. (1983). *J. Polym. Sci. Polym. Phys. Ed.*, **21**, 2027.
70. BLUMSTEIN, A., BLUMSTEIN, R. B., GAUTIER, M. M., THOMAS, O. and ASRAR, J. (1983). *Mol. Cryst. Liq. Cryst. Lett.*, **92**, 87.
71. WINDLE, A. H., VINEY, C., GOLOMBOK, R., DONALD, A. M. and MITCHELL, G. R. (1985). *Faraday Discuss. Chem. Soc.*, **79**, 55.

*Chapter 4*

# STRUCTURAL STUDIES OF FIBRES OBTAINED FROM LYOTROPIC LIQUID CRYSTALS AND MESOPHASE PITCH

M. G. Dobb and D. J. Johnson

*Textile Physics Laboratory, Department of Textile Industries,
University of Leeds, UK*

## 1. INTRODUCTION

The ever demanding requirements of modern industry have necessitated the development of new materials of high mechanical performance. In particular the need for composite components and reinforcing elements of improved specifications in high technology engineering has provided the impetus for the relatively recent rapid advances in fibre physics and technology.

Although the observed strengths and moduli of the commercially available fibres are still somewhat lower than the theoretical estimates, it is encouraging to note that, with the gradual elucidation and subsequent elimination of performance-limiting structural defects, the actual tensile parameters of some fibres are now closely approaching the theoretical values.

Examination of the advances clearly shows the adoption of three essentially distinct approaches to the problem of producing fibres with exceptional mechanical performance such as high modulus and high strength, namely:

1.  The production of straight chain molecular arrays and the achievement of high orientation in simple polymers such as polyethylene by 'superdrawing' of melt-spun or gel-spun fibres.

2.  The formation of carbon layer planes, and the attainment of highly aligned layer-plane structures, to produce high performance carbon fibres by the oxidation and carbonization under tension of fibres such as those based on poly(acrylonitrile).

3.  The selection of suitable high polymer systems capable of forming nematic liquid-crystalline phases, for subsequent extrusion of highly aligned molecular domains into fibres without the necessity of subsequent conventional drawing processes.

In this chapter we shall confine our attention to this latter approach and consider in particular the development and structure/property relationships in such fibres as the PPT aramid Kevlar obtained via a lyotropic mesophase and carbon fibres derived from thermotropic mesophase pitch.

## 2.  POLY(p-PHENYLENE TEREPHTHALAMIDE) FIBRES

### 2.1. Production

Although the predecessors of Kevlar type fibres were based on poly(p-benzamide),[1] the present versions are known to consist essentially of poly(p-phenylene terephthalamide) (PPT). Synthesis involves a condensation reaction of p-phenylenediamine with terephthaloyl chloride:[2]

A mixture of hexamethylphosphoramide and N-methylpyrrolidone is used to dissolve the p-phenylenediamine component and the terephthaloyl chloride added subsequently in a nitrogen atmosphere.[3] The use of the potentially carcinogenic hexamethylphosphoramide has caused some concern, and as a result alternative routes have been devised.[4]

The presence of impurities, including traces of water in the monomers and solvent, tends to impair the formation of high molecular weight polymer. Removal of solvent and hydrogen chloride from the condensation product is also highly desirable.

### 2.1.1. Fibre Formation

Perhaps one of the most important characteristics of *para* aromatic polyamides from the viewpoint of fibre formation is their ability, under certain conditions of concentration, type of solvent, and temperature, to

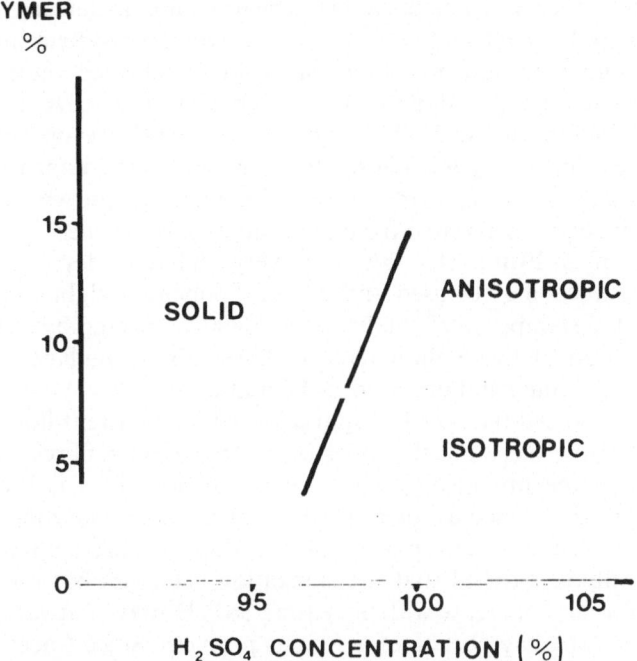

FIG. 1.    Phase diagram for solutions of PPT in sulphuric acid.

form liquid crystalline solutions[1,3] which are nematic in character. A typical phase diagram is shown in Fig. 1. Such solutions exhibit an unusual relationship between viscosity and polymer concentration. As with isotropic polymer solutions, the viscosity initially increases with increasing concentration; however, at a critical point an anisotropic phase separates and the viscosity falls abruptly as additional polymer is added. Such behaviour is particularly useful since it allows the preparation of dopes containing relatively high proportions of polymer suitable for spinning.

Among the properties relevant to the production of fibres of high performance is the response of the nematic phase to shearing. Under conditions of extensional flow such as those encountered during extrusion, the degree of parallelization of nematic domains becomes so high that the product is already highly oriented without the necessity of a subsequent drawing process, as is customary in processing conventional fibres initially of low orientation. Thus PPT fibres tend to possess superior tensile properties to those exhibited by conventional fibres.

Unlike thermotropic polymers, which require no solvent for formation of a liquid-crystal phase, PPT forms a lyotropic system and cannot be melt-spun. Consequently, fibre formation is achieved via a wet-spinning process involving a dope consisting of PPT in a solvent of preferably 99·8% sulphuric acid. Although conventional wet-spinning methods produce fibres of good tensile performance, much improved properties can be achieved using a dry-jet wet-spinning technique whereby the face of the spinneret is separated from the coagulant by an inert fluid layer of the order of 5–20 mm. In this way very different dope and coagulant temperatures may be used. Indeed, for a polymer weight content of about 20%, dope temperatures in the range 70–90°C are employed, whilst for the production of the highest tenacity fibres, the coagulant liquid, usually water or dilute sulphuric acid, is held at about 5°C.

It is reported that vertical spin tubes with co-current flow of coagulant improve the mechanical properties of the as-spun fibres. An important factor in the processing of high performance fibres is the spin-stretch factor (SSF) defined as the ratio of the fibre velocity leaving the coagulant bath to the average velocity of the dope in the spinneret capillary. Generally it is found that the elongation decreases but the tenacity and modulus both increase with increasing SSF. However, at values around 14, filament breakage becomes a serious problem. Since traces of acid have adverse effects on tenacity, it is usual to spray the threadline with a weakly alkaline solution before final washing in water and subsequent drying.

A vital factor necessary for the development of high tensile modulus is the thermal treatment of the as-spun fibres. Essentially, tensioned fibres are passed through a heated tube in an inert atmosphere at zone temperatures between 450 and 550°C for 0·5 to a few seconds. More prolonged heating above 450°C leads to excessive degradation and consequent diminution in performance. It is probable that suitable thermal treatments promote molecular chain displacements, thus permitting the formation of regular bonding between the highly oriented regions of crystalline order. This short heat treatment is to be contrasted with the extensive times (of the order of hours) necessary to promote significant increases in modulus of as-spun thermotropic fibres.[5] In the latter case it is likely that the improved performance results from increases in molecular weight of the polymer and changes in crystal structure.

## 2.2. Structure
A distinctive feature of the PPT molecule is its essential linearity, which is derived from the bonding of rigid phenylene rings in the *para* position.

This is to be contrasted with the irregular chain conformation of Nomex fibres, poly(*m*-phenylene isophthalamide), where the phenylene and amide units are linked in the *meta* position. Clearly, from the viewpoint of packing efficiency, the reduction in the effective cross-sectional area occupied by a single chain, and the probability of a higher orientation, would be expected to lead to PPT exhibiting superior tensile properties.

### 2.2.1. Crystallography

The occurrence of amide groups at regular intervals along the linear chains should facilitate extensive intermolecular hydrogen bonding and lead to a high degree of order or crystallinity. Indeed, confirmation is provided by wide-angle X-ray and electron-diffraction patterns of Kevlar fibres which reveal a considerable number of discrete reflections and layer lines, as shown in Fig. 2.

It is interesting to note that PPT is polymorphic, the type of crystal modification obtained being dependent on the polymer concentration in the dope used for spinning.[6] When the polymer is spun from highly

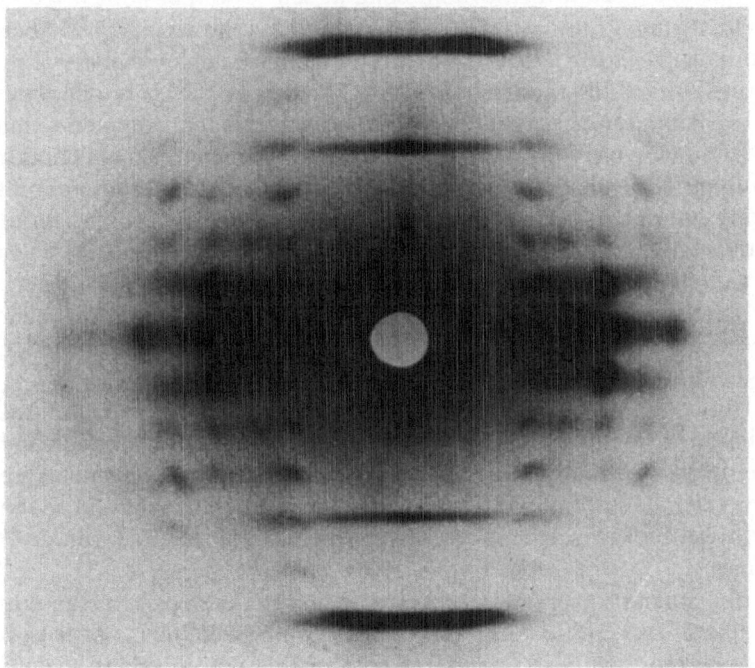

FIG. 2.   Wide-angle X-ray diffraction pattern of Kevlar 49.

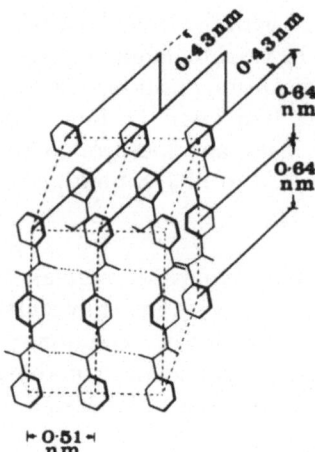

FIG. 3.    Unit cell of one crystallographic form of PPT.

concentrated anisotropic solutions, the chains form an essentially monoclinic (pseudo-orthorhombic) unit cell[7] of dimensions $a = 0.719$ nm, $b = 0.518$ nm, $c$ (fibre axis) $= 1.29$ nm, $\gamma = 90°$, as shown in Fig. 3. There are two molecules per unit cell and two monomeric units in the axial repeat, giving a crystallographic density of $1480 \, \mathrm{kg \, m^{-3}}$. Fibres derived from dopes containing relatively low concentrations of polymer exhibit a different packing equivalent to a lateral displacement $(b/2)$ of chains along alternate (200) planes. Both crystal modifications are present in fibres spun from intermediate polymer concentrations. The most favourable symmetry for the PPT crystal should be $P2_{1/n}$ but, because of a lack of perfect two-fold screw symmetry, this is reduced to $P_n$. Furthermore, Northolt[7] suggests that the orientation angles between the phenylene planes and the amide group are 38° and $-30°$ for the $p$-phenylenediamine segment and the terephthalic segment respectively.

From a rigidity viewpoint, bond rotation and hence molecular flexibility is inhibited by the presence not only of the aromatic rings but also by the double-bond nature of the amide group arising from resonance effects. Moreover, lateral interactions between adjacent chains are in the form of hydrogen bonds restricted to the (200) lattice planes and van der Waals forces.

The apparent average lateral crystallite size derived from the resolved profiles of the 110 and 200 X-ray diffraction reflections is of the order of 5–6 nm. This value is considerably less than the correlation length of 20–100 nm obtained from analysis of the axial 006 reflections.

Despite the general sensitivity of organic polymers to electron irradiation (manifest by the decay of electron diffraction patterns), lattice-fringe images derived from 110 equatorial reflections at 0·433 nm and 002 meridional reflections at 0·645 nm have been recorded from PPT,[8] as shown in Fig. 4. It is clear that the equatorial arrays are considerably less extensive than the meridional fringes and show little evidence of lattice distortion apart from the appearance of curved layer planes particularly in the meridional arrays. Using the lattice-fringe technique, data have also been reported regarding the distribution of crystallite widths.[8] Thus in Kevlar 49 it is rare for the crystallite width to exceed 9·0 nm.

Although it is claimed[9] that soft X-rays such as $AlK_\alpha$ produce a weak meridional reflection at about 30 nm (see Section 2.2.5), there is no evidence for the small-angle two- or four-point meridional reflections traditionally associated with chain folding in conventional polyamides, using $CuK_\alpha$ radiation. Moreover, the series of meridional spacings derived from X-ray wide-angle studies are compatible with an extended chain conformation in PPT fibres.

### 2.2.2. Radial Orientation

Unlike conventional textile fibres which do not generally exhibit any preferred lateral orientation, PPT fibres such as Kevlar, and carbon fibres from mesophase pitch, display radial or circumferential orientation. For example, cross-sections of Kevlar (2 μm thickness) have a characteristic Maltese cross appearance when viewed between crossed polars in a light microscope. The unavoidable distortions associated with cutting thin transverse sections of Kevlar suitable for electron microscopy have necessitated the study of either oblique or longitudinal sections. Nevertheless, electron diffraction and transmission electron microscope dark- and bright-field studies confirm the unusual radial orientation.[10,11] It is worth noting, however, that electron diffraction patterns of longitudinal sections taken from different regions of a fibre always show both main equatorial 110 and 200 reflections although their relative intensities change. Radial orientation cannot therefore be considered to be perfectly developed, and the radial sheets should be regarded as an aggregation of similarly oriented crystallites, although exhibiting a restricted distribution of orientations of the hydrogen-bonded planes.

### 2.2.3. Pleat Structure

Strong evidence for an unusual supramolecular axial organization was first detected in longitudinal sections of commercially available PPT fibres

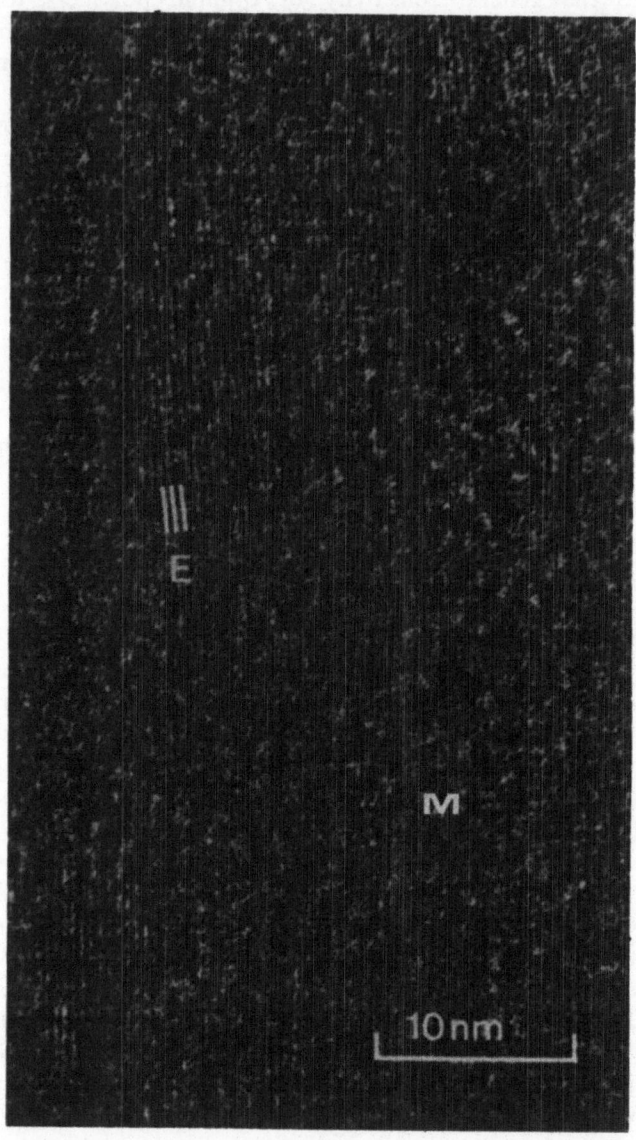

FIG. 4. Transmission electron microscope lattice fringes derived from 0·433 nm equatorial reflections (E) and 0·645 nm meridional reflections (M).

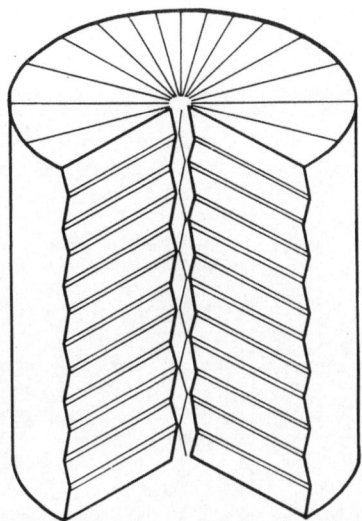

FIG. 5.   Schematic model of radial pleated-sheet structure of a PPT fibre.

such as Kevlar 49 and Kevlar 29 from the analysis of dark-field TEM images.[10] Thus a system of narrow bands 30 nm in width spaced at intervals of 250 nm is associated with the meridional 006 reflections, whilst another system of broad bands of spacing approximately 500 nm is associated with the off-meridional reflections. Detailed examination involving tilting of the sections indicated a uniform distribution of ordered material throughout the fibre and that the banding resulted from periodic changes in crystalline orientation. Similar banded structures have since been observed in some thermotropic fibres.[12]

It has been suggested that the pleating results from the viscoelastic behaviour of liquid crystals which shrink elastically after release of stretching during fibre formation.[13] Because polymer liquid crystals have sufficiently long relaxation times, the zig-zag structures produced in the dope by shrinkage remain in the fibres after solidification.

It is now generally accepted that the fibres are composed essentially of radially oriented axially pleated sheets as shown in Fig. 5, with the alternating components of each sheet arranged at approximately equal but opposite angles, and that only over a short transitional band (∼ 30 nm) are the PPT molecules parallel to the fibre axis. Measurements of the angular separation of the peaks in the 200 electron diffraction arc indicate that the angle between adjacent components of the pleat is about 170°.

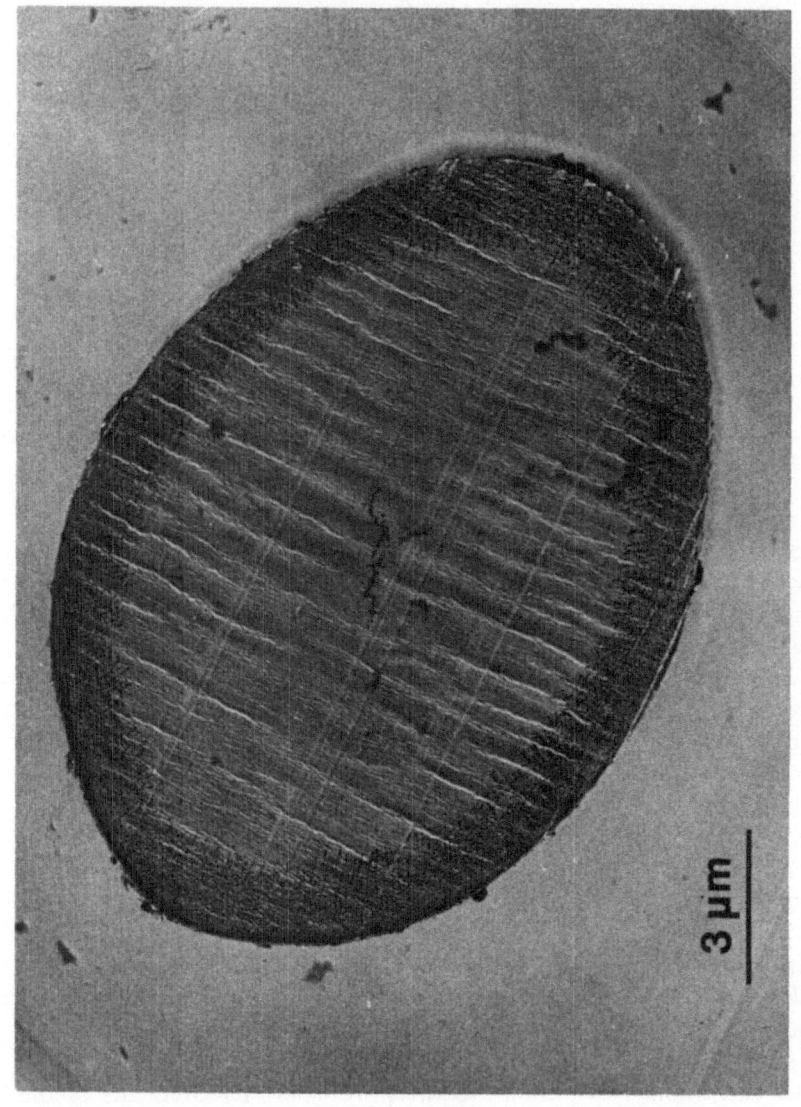

FIG. 6. Bright-field electron micrograph of an oblique section through a silver sulphide stained Kevlar 49 fibre showing skin—core organization.

The persistence and hence mechanical stability of the pleat, at least during the short-term application of stress up to the point of fibre fracture, has been demonstrated directly in video-tape sequences using a specially designed stretching device in a scanning electron microscope.[14] Furthermore it would appear that the pleat is relatively stable to moderate temperatures although heat treatment around 500°C for a few minutes causes exaggerated pleating (i.e. progressive reduction in the angle between adjacent components) prior to degradation.[10]

### 2.2.4. Skin–Core Differentiation

Dry PPT fibres usually exhibit diffuse but intense equatorial scattering in the X-ray small-angle pattern[15] similar to that exhibited by carbon fibres. It is pertinent to note that the scattering may be modified by appropriate treatment; for example, under moist conditions a significant reduction is observed. Alternatively, if Kevlar fibres are subjected to gaseous hydrogen sulphide and subsequently treated with silver nitrate to form silver sulphide[11] (a relatively good scatterer of X-rays), then the intensity increases. Such observations are compatible with the presence of electron density discontinuities such as voids within the fibre which are accessible to reagents. X-ray data and electron microscope imaging of the silver sulphide deposits both indicate that the voids are needle-shaped, having lengths of about 25 nm and widths in the range 5–10 nm, oriented roughly parallel to the fibre axis. Moreover, electron microscope studies of silver sulphide labelled fibres strongly suggest that the microvoids occur predominantly in a layer around the fibre periphery,[15] as shown in Fig. 6. On this basis, therefore, it is possible to define a relatively thin (0·5–2·0 μm) porous skin surrounding a more compact core region. There is also evidence showing a tendency for the voids to be located along radii of the fibres.

On the basis of dye absorption on damaged or staple fibres, some workers[16] have claimed that the core is more porous than the skin, and that the silver sulphide used in other studies to reveal the location of voids has penetrated only the skin and not the porous core. However, since large deposits of silver sulphide up to 16 μm in length and 0·08 μm in width (associated with macrocracks) are frequently found in the fibre core in addition to the peripheral layer (corresponding to the skin region), then the penetration argument becomes somewhat tenuous. Indeed, the preferential localization of dye observed by Panar et al.[16] within the core adjacent to the cut ends of the fibre most likely results from the formation of interior cracks associated with the mechanical damage induced during cutting operations. On balance, therefore, it would seem reasonable to accept a highly porous skin of variable dimensions (up to about $2 \mu m$ in thickness) depending on the coagulation conditions, with a generally less porous core containing large axial cracks.

### 2.2.5. Defect Layers

Morgan et al.[17] consider that the chain-end concentration and their distribution within the fibre are primary critical physical structural parameters that control deformation and failure. Indeed, from fracture topography observations of both virgin and acid-etched fibres, they concluded that PPT molecules in the skin are arranged differently from those in the core. From such studies a model of the chain-end distribution in PPT fibres was formulated in which the chain ends in the fibre exterior (skin) are arranged essentially randomly relative to one another but become progressively more clustered in the fibre interior (core) resulting in periodic transverse weak planes every 200 nm, the average PPT macromolecular length along the fibre axis. Although such a proposal has attractive features, it is difficult to reconcile with the fact that fibres with significantly different skin thicknesses generally exhibit very similar tensile strengths. As such, therefore, the model is unlikely to account fully for the failure mechanism.

On the other hand, Panar et al.[16] have considered a combination of morphological aspects which are thought to determine the unique mechanical properties of PPT fibres. In addition to the pleat structure these include (a) defect layers, (b) fibrillar character, and (c) skin–core differentiation. They report that ordered layers of axial periodicity 30–40 nm can be observed in transmission electron micrographs of longitudinal sections of undegraded material which can be enhanced by both chemical and plasma etching techniques. Such organization has been supported by the detection of weak meridional small-angle X-ray scattering at about 30 nm with $AlK_{\alpha}$ radiation. Panar et al. suggest that this periodicity represents a potentially weak link in the structure which must be successfully bridged to result in satisfactory strength, and that the manner in which the defect zone is bridged distinguishes PPT from conventional fibres. Indeed, the axial correlation length (about 80 nm) is much larger than the defect periodicity. This is in marked contrast to that found in conventional fibres where it will be recalled that the crystalline correlation length (crystal size) is smaller than the long period. The observations are consistent with a defect layer which includes possibly 50% of chains ends. The increased length over which register is maintained may be due to a high proportion of chains being extended as they pass through the defect bands. Such a concept is similar to that originally proposed by Ward and coworkers[18] to account for the mechanical behaviour of high-modulus polyethylene fibres.

### 2.2.6. Fibrillar Character

There are many references in the literature to the formation of fibrils approximately 600 nm across on fracturing PPT fibres. However, whether the fibres consist of fibrils or whether these are simply the result of cleavage along the c-axis of the crystalline polymer remains an open question despite attempts to resolve the problem. Certainly PPT fibres are highly crystalline and there is no evidence for a sizeable disordered fraction akin to a matrix. Moreover it should be noted that the quoted lateral dimension of the fibrils is approximately one-hundred times that of the average lateral crystallite size.

### 2.3. Mechanical Behaviour

As might be expected, orientation of the main chains about the fibre axis plays a major role in determining the mechanical performance of fibres. For example, Carter and Schenk,[19] using a compilation of data from various sources, showed that the initial modulus of PPT and poly(p-benzamide) fibres increased in an essentially exponential manner with decreasing orientation angle.

More recently, specific studies on specially selected spools of commercial Kevlar 49 showed that the half-height azimuthal width of the 004 X-ray reflection decreases almost linearly with both increasing modulus (over the range 117–147 GPa) and lateral crystalline size.[14] Indeed it would appear that the better the orientation the greater the crystallite size.

Further work on these same specimens indicated that the skin/core ratio may also contribute to differences in modulus. Thus the lower-modulus fibres always exhibited a thicker skin (as revealed by silver sulphide deposition) than the higher-modulus specimens. No doubt the higher void content associated with a thick skin implies that the number of polymer chains per fibre cross-section is less than in corresponding fibres exhibiting a thin skin, in which case, other things being equal, one would expect to observe a lower tensile modulus.

### 2.3.1. Tensile Properties

The tensile parameters summarized in Table 1 illustrate the exceptionally high performance characteristics for an organic polymer exhibited by

## TABLE 1
### TENSILE PROPERTIES OF HIGH-PERFORMANCE FIBRES

| | Kevlar 29 | Kevlar 49 | Nomex | Carbon | | | |
| | | | | Pan-based | | | Mesophase pitch based |
| | | | | type I | type II | type A | |
|---|---|---|---|---|---|---|---|
| Density (kg m$^{-3}$) | 1 440 | 1 450 | 1 400 | 1 950 | 1 750 | 1 760 | 2 220 |
| Tensile strength (GPa) | 2·64 | 2·64 | 0·7 | 1·8–2·2 | 3·1–4·4 | 2·9–4·3 | 1·4–2·2 |
| Modulus (Gpa) | 58·9 | 127·5 | 17·3 | 340–517 | 230–276 | 230–240 | 140–820 |
| Elongation at break (%) | 4·0 | 2·4 | 22·0 | 0·4–0·6 | 1·3–1·6 | 1·3–1·8 | 0·2–1·0 |

Kevlar fibres. Since the theoretical strength and modulus of PPT calculated from bond dissociation energy, force constants, and crystallographic data are about 20 GPa and 200 GPa respectively,[20] it should be noted that the observed properties provide the closest approach to theoretical values of any commercial organic fibre, particularly with respect to modulus. The combination of low density with high strength and modulus gives Kevlar one of the highest specific tensile strengths of any commercially available material, and a reasonably high specific modulus even when compared with carbon fibres.

The stress–strain curve in the case of Kevlar 49 is almost linear, and Northolt and Van Aartsen[21] concluded that, for a large part of the curve, the macroscopic strain is mainly brought about by the elongation of the crystal lattice through valence angle deformation and bond stretching of the polymer chain. They suggest that the deviation from linearity which starts near 1% strain could be attributed to chain breakage and other irreversible processes such as crystalline rotation towards the fibre axis.

Recently, Northolt and Hout[22] have questioned whether previously developed deformation theories for semi-crystalline polymers provide a proper description for the tensile deformation of wholly paracrystalline fibres composed of rigid-rod chains such as PPT with a narrow orientation distribution. As a result they have presented a new mathematical derivation for the stress–strain curve and have shown that the shape of the initial crystallite orientation distribution and the modulus for shear parallel to the chain direction are important factors determining the stress build-up during the extension of PPT fibres. The new equation for the extensional compliance ($s_{33}$) does not contain the fourth moment $\langle \sin^4 \theta \rangle$

of the orientation distribution as for the classical series aggregate model, but is of the form

$$s_{33} \simeq \frac{1}{e_3} + \frac{\langle \sin^2 \theta \rangle}{2g}$$

where $e_3$ is the chain modulus and $g$ is the shear modulus.

In cyclic loading, Kevlar 49 appears to be essentially elastic after the first loading cycle and shows a narrow hysteresis loop. Bunsell[23] considers that the high loads and large number of cycles to which the fibre has to be cycled for fatigue failure to occur mean that fatigue of the individual fibres in a composite should not be a serious hazard.

A feature, highly desirable in many engineering applications, is the remarkably low creep characteristic. For example, at room temperature with a loading of 50% of ultimate tensile strength, the creep after a brief period of initial growth is of the order of 2·5%. Kevlar 49 exhibits a short-term stress relaxation which for the stress range 0·14–1·0 GPa amounts to a relaxation of some 6–8% in the time interval 0·1–300 s.[24] Subsequent creep is generally linear with log time. Although tensile strength is only slightly affected by water immersion, the effect on creep is more marked. A sharp discontinuity is seen in the creep curve following water immersion, though this increment of creep also proves approximately linear with log time taken after immersions. Cook et al.[24] suggest that not only is creep rate increased by water but also the number of creep sites is increased. They also consider that creep does not have a great influence on stress–rupture behaviour which is thought to be associated with fracture of polymer main chains. In stress–rupture studies the influence of oxygen appears to be small and it is found that the mean log lifetime is approximately related to the square of the stress. If a first power of stress is used, then linear extrapolation to low residual strength predicts, surprisingly, geological lifetimes for Kevlar 49 at ambient temperatures.

The tensile fracture morphology of Kevlar is complex and when it is broken in air it shows evidence of jagged faces and severe axial splitting over a considerable fibre length; this is in marked contrast to the almost perfect transverse fracture of carbon fibre. It is interesting to note that, when fractured in glycerol or other viscous media, the fracture faces are generally smooth, long, and wedge shaped, often revealing the internal pleat structure.[25] It is likely therefore that the energy released during fracture in air produces further damage around the initial region of failure, whereas the presence of a viscous medium tends to limit the damage. The insensitivity of Kevlar to notches in the surface layer is also worthy of mention which, unlike e.g. glass fibres, is compatible with the concept of weak interchain bonding allowing deflection of an originally transverse crack along a direction parallel to the fibre axis.

### 2.3.2. Compressive Behaviour

Although PPT fibres such as Kevlar exhibit exceptionally high tensile properties, their compressional behaviour is somewhat disappointing.[26] Information about this particular aspect has been obtained from either fibre bending or axial compression of Kevlar-reinforced composites.

In simple bending tests, a characteristic feature is the formation of oblique kink bands where compression is limited to the innermost region of the bent fibres. The detailed mechanism of the deformation has been reported by Dobb et al.[27] It would appear that the onset of plastic deformation during compression arises from abrupt localized changes in orientation of the molecular chains to form kink bands (i.e. all the chains within the kink band will be sheared with respect to each other). Further compression produces a pile-up of kink bands which result in the appearance of bands of extruded material in the fibre surface.

Fibres extracted from a composite system whose matrix undergoes shrinkage (compressive strain $< 3\%$) exhibit bands around the surface at both $+45°$ and $-45°$ to the fibre axis.[28] Under uniaxial compression the maximum shear stress in a fibre occurs at $45°$ to the axis; consequently the observed kinking suggests that the fibres yield in a shear mode at small compressive strain, $< 3\%$. The compressed fibres show an approximately 20% loss in strength and a variable but significant loss in modulus. This behaviour implies that there is a straightening out of regions which buckled as a result of compression when subsequently stressed. In contrast to carbon fibres, the aramids can survive bending intact by yielding in axial compression;[29] this is important technologically in terms of converting fibres into complex textile assemblies such as fabrics, by say the weaving process.

Little information is available about the transverse compressive behaviour, although the work of Phoenix and Skelton[30] suggests that Kevlar exhibits substantial plastic flow, which is in contrast to the case of carbon fibres where the deformation is almost completely elastic up to the point of brittle failure. Reported values for the transverse compressive modulus of Kevlar 49 and 29 are 0·76 and 0·77 GPa respectively.[31]

The fundamental reason for the relatively poor compressional behaviour undoubtedly resides in the shearing of molecular chains due to the weak lateral bonding involving oriented hydrogen bonds between the polymer chains. It will be recalled that in other competitive materials, such as PAN-based carbon fibre, which possess significantly higher compressional strength, the interatomic bonds in the carbon layers are strongly covalent, and the layers are more randomly oriented.

## 3.    MESOPHASE PITCH BASED CARBON FIBRES

### 3.1. Production
Many different precursor materials have been used in the production of high-performance carbon fibres, but only poly(acrylonitrile) (PAN) and pitch have any real commercial significance. PAN-based fibres have been the market leaders, but pitch has long been heralded as a low-cost raw material which could make carbon fibres more cost-effective for many varied end-uses.[32] Early work on isotropic pitches by Otani and his coworkers[33-36] led to the production of petroleum pitch-based carbon fibres by the Kureha Kagaku Company.[37] The process included melt spinning into fibre form, oxidation to render the fibres infusible, and carbonization at 1000°C. In order to obtain high preferred orientation, which is essential for high modulus, hot stretching at temperatures in excess of 2500°C is required;[38] this is a difficult and costly procedure which eliminates the price advantage of pitch.

A crucial step towards a cheaper and more convenient method for producing carbon fibres from pitch was the discovery, by Brooks and Taylor,[39,40] of a carbonaceous mesophase in the pitch to coke conversion. The formation of mesophase globules within the body of a molten pitch has been described by Barr et al.[41] When pitches which contain a certain mixture of aromatic compounds are heated above 350°C, condensation reactions form large molecules which aggregate into a liquid crystalline phase with nematic order. This 'mesophase' has a higher surface tension than the low molecular weight isotropic liquid phase; the globules grow in size, coalesce into large spheres, and eventually form extended anisotropic regions.

Singer[42] has described experiments in which mesophase fibres drawn from acenaphthylene pitch at 450°C could be heat treated without any oxidative thermosetting step, and were anisotropic with a high degree of axial preferred orientation. Further experiments led to Union Carbide's mesophase pitch based carbon fibre process.[43] The most suitable pitches were shown to be those capable of producing large domain mesophase and in which there is more than 40% mesophase in the two-phase system.

Diefendorf[44] has summarized the three major requirements of a mesophase pitch as (1) thermally stable at spinning temperatures, (2) free from particulates, (3) rheologically acceptable. Riggs[45] has discussed the general characteristics of pitches and shows how the solvent extraction of a pitch with a solvent system having the appropriate solubility parameter can give a more or less ideal fraction for carbon fibre production.

## 3.2. Structure

The first detailed account of mesophase pitch based carbon fibres reported that there were at least three observable macrostructures: radial, circumferential (or tangential), and random.[42] These structures were present in the as-spun fibres, according to polarized-light micrographs, and were retained through carbonization, as was evident from SEM images. X-ray diffraction showed that, in the most highly oriented fibres, the layer-plane stacks which make up the macrostructure are highly crystalline with several *hkl* lines, thus indicating the presence of 3D graphite.

We may note that, in general, carbon fibres do not have a regular arrangement of the layer planes and are said to consist of turbostratic graphite. Certainly it is only after exceptional treatments, for example catalytic graphitization, that PAN-based carbon fibres show even the smallest traces of 3D graphite.[46] Normally PAN-based carbon fibres are considered non-graphitizing. The hexagonal unit cell of graphite, illustrated in Fig. 7, has dimensions $a = b = 0.246$ nm, $c = 0.6707$ nm,

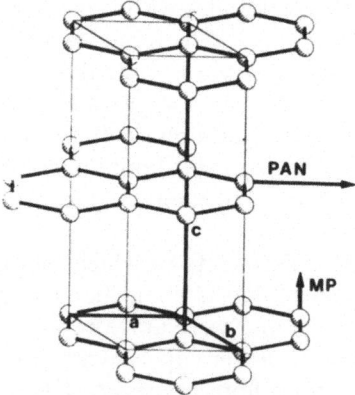

Fig. 7.   Unit cell of graphite; arrows indicate direction of fibre axis in mesophase pitch based carbon fibres (MP) and PAN-based carbon fibres (PAN).

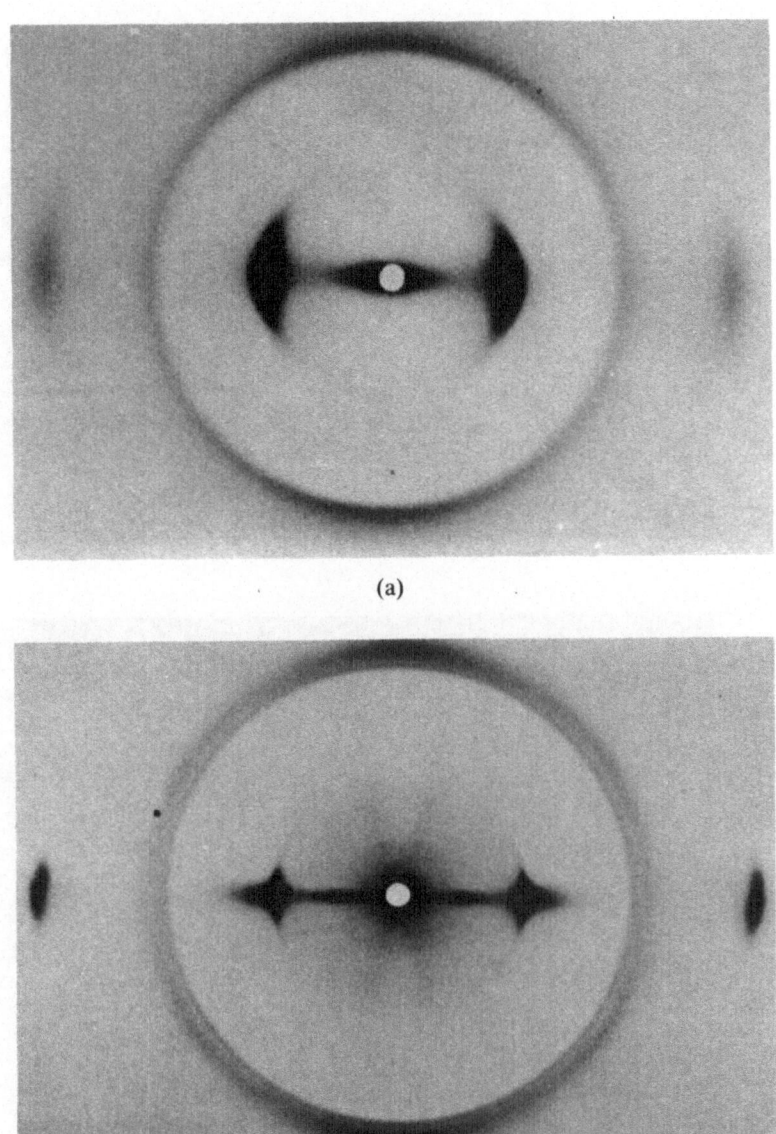

(a)

(b)

FIG. 8.   Wide-angle X-ray diffraction patterns of (a) P-100, (b) P-120.

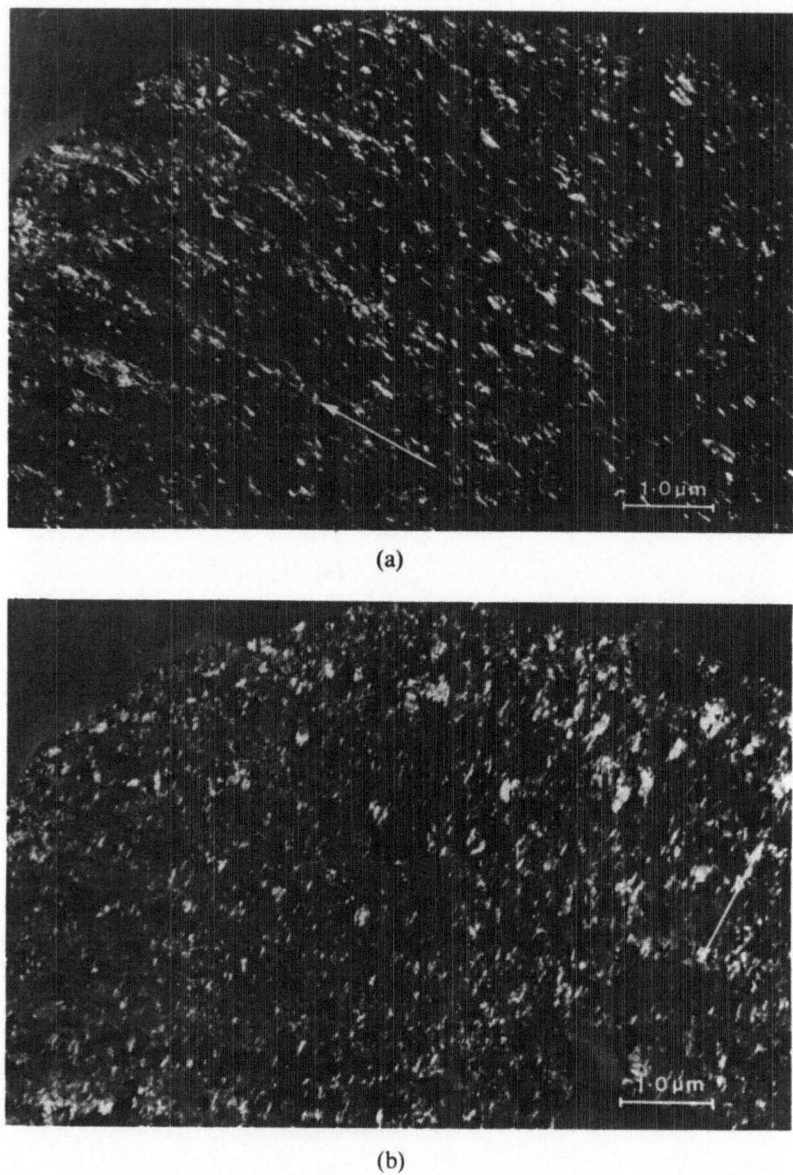

(a)

(b)

FIG. 9.  Dark-field (002) micrographs of transverse section of mesophase pitch based carbon fibre, with selected area aperture in two positions at 90° to each other; arrows show radial organization of crystallites.

giving an interlayer spacing $c/2$ of 0·3354 nm. The values of the interlayer spacing for the Union Carbide range of mesophase pitch based carbon fibres vary from 0·346 nm for P-25 to 0·337 nm for P-120.[47] Figure 8 illustrates X-ray diffraction patterns in (a) P-100 fibres and (b) P-120 fibres. The latter has a 101 ring of stronger intensity than the 100 ring, clear evidence for very highly graphitized material; a hexagonal distribution of intensity is also indicated around these rings. This $a$-axis orientation of the layer planes has been discussed in detail by Ruland and Plaetschke.[48] The arrows in Fig. 7 indicate the direction of the fibre axis in the case of both PAN and mesophase pitch based carbon fibres.

Electron spin resonance measurements showed a much greater $g$ anisotropy than for PAN-based carbon fibres, which is another indicator of the greater graphitizability of the mesophase pitch fibres. Other workers, for example Chwastiak et al.,[49] observed circular section fibres with sheet-like structures which were neither entirely radial nor entirely circumferential. Radial organization (arrowed) is clearly visible in dark-field (002) images (Fig. 9) from a transverse section of a mesophase pitch based carbon fibre with the selected-area aperture placed in two positions 90° apart on the 002 diffraction ring. A dark-field (002) image from a longitudinal section of P-100 mesophase pitch based carbon fibre is shown in Fig. 10, and a typical lattice-fringe image from the same section is illustrated in Fig. 11.

White et al.[50] have reported mesophase pitch based carbon fibres with radial and random structures; they describe two often observed examples, the radial structure with a pie-shaped missing wedge, and the mixed structure with an outer radial section and an inner random section. They also demonstrated that variable structure along the length of the fibre can be observed in many cases. More importantly, the different structures and the changes between them were discussed in terms of disclinations.

### 3.2.1. Disclinations

It is well known that real crystals are never perfect; they contain imperfections which govern the properties of the material. In metals and semiconductors, translations or linear displacements of one part of a crystal with respect to another give rise to dislocations. In liquid crystals, if part of a solid is displaced by a rotation rather than a translation, the result is a disclination.[51] Figure 12 shows examples of positive and negative wedge disclinations of the type found in mesophase carbonaceous systems, as explained in detail by Zimmer and White.[52] Deformation studies showed that a dense array of parallel wedge

FIG. 10. Dark-field (002) micrograph of longitudinal section of P-100 fibre.

disclinations forms when plastic mesophase is drawn to a fine fibre. As long as the mesophase remains sufficiently fluid for disclination motion, annihilation reactions occur between adjacent disclinations of opposite sign; when cooling takes place after spinning, the bulk of the fibre consists of parallel wedge disclinations of opposite sign. Under certain processing conditions there remains only a single positive disclination and a radial structure results, with a missing wedge as a natural crystallization

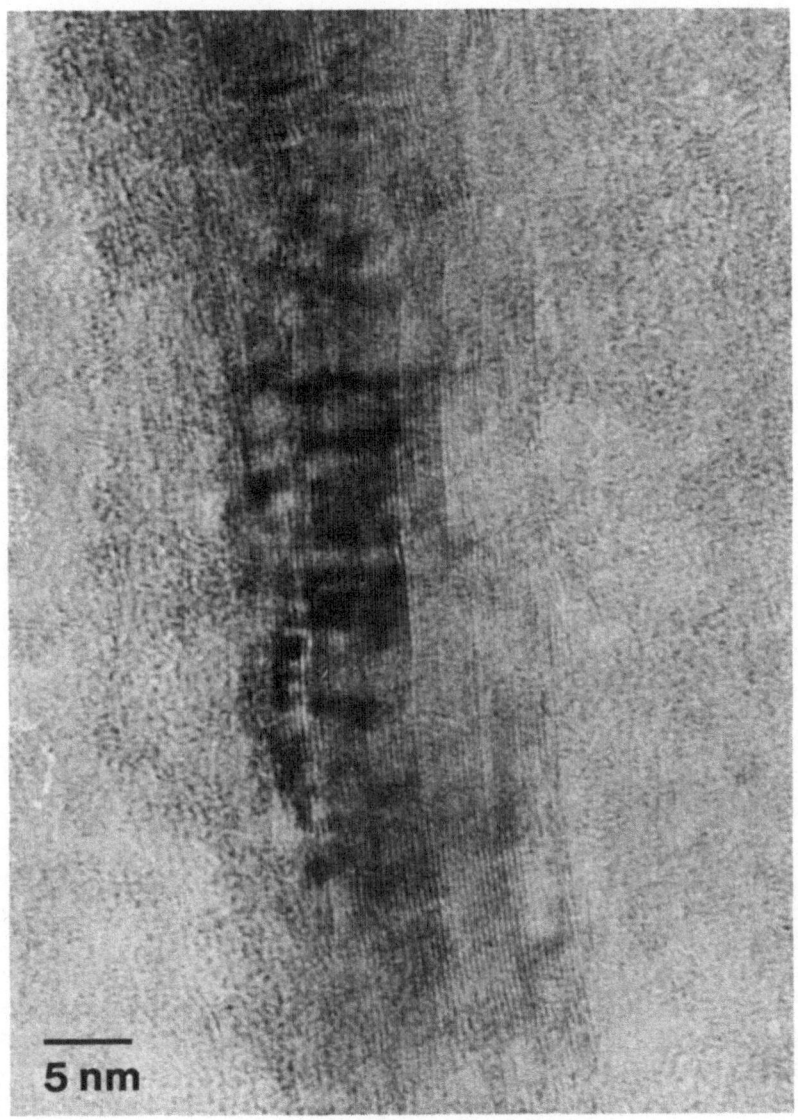

5 nm

FIG. 11.   Lattice fringe image of longitudinal structure of P-100 fibre.

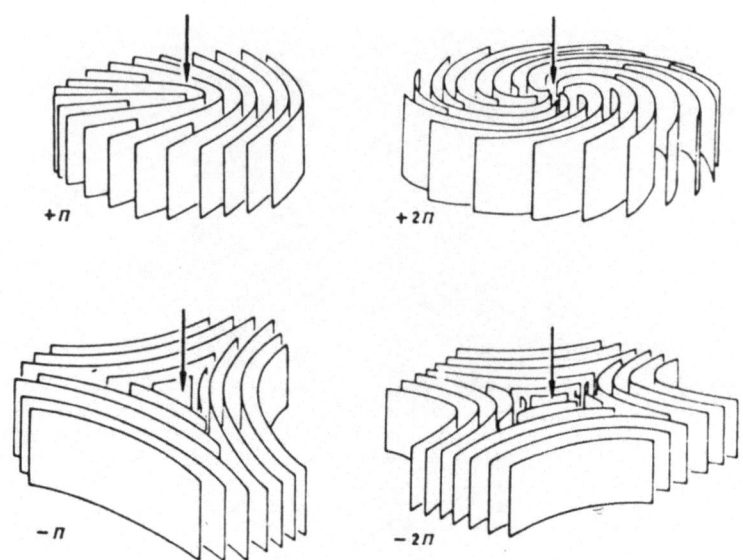

FIG. 12.    Wedge disclinations.

phenomenon in some cases. The oriented core type structure is another natural consequence of crystallization from the mesophase.[53]

### 3.3. Mechanical Behaviour

The isotropic pitch based carbon fibres produced by Hawthorne[38,54] had a Young's modulus ($E$) of 40 GPa and tensile strength ($\sigma$) of 0·70 GPa, very similar to the values for a commercial Kureha fibre (see Table 2). After hot stretching, Young's moduli of up to 600 GPa could be achieved at a tensile strength of about 2 GPa,[54] whereas the hot stretched Kureha fibres reached a more modest 100 GPa modulus and 1·39 GPa in tensile strength. The first commercially produced mesophase pitch based carbon fibres were described by Volk of Union Carbide in 1974.[32] They were known as Thornel type P, the continuous filament type having a Young's modulus of 240 GPa and a tensile strength of 3·10 GPa. When the yarn was processed at higher temperatures, a tensile modulus of 620 GPa could be achieved with a tensile strength of 1·55 GPa. These tensile strengths were determined from short gauge-length measurements on single filaments; manufacturers' data later showed that production yarn had an average tensile strength of 1·39 GPa at a modulus of 345 GPa. The current range of Thornel fibres, P-25 to P-120, have the tensile properties given in Table 2.

The theoretical modulus of a carbon fibre composed of perfect layer

TABLE 2

TENSILE PROPERTIES OF PITCH-BASED CARBON FIBRES

| Manufacturer | Fibre | $E$ (GPa) | $\sigma$ (GPa) | $e$ (GPa) |
|---|---|---|---|---|
| Hawthorne | Isotropic as-spun | 40 | 0·70 | — |
| | Hot stretched | 600 | 2·00 | — |
| Kureha | Isotropic as-spun | 30 | 0·77 | — |
| | Hot stretched | 100 | 1·39 | — |
| Union Carbide | Early production | 345 | 1·39 | — |
| Union Carbide | Thornel P-25 | 140 | 1·40 | 1·0 |
| | P-55 | 380 | 2·10 | 0·5 |
| | P-75 | 500 | 2·00 | 0·4 |
| | P-100 | 690 | 2·20 | 0·3 |
| | P-120 | 820 | 2·20 | 0·2 |

$E$, Young's modulus; $\sigma$, tensile strength; $e$, strain to failure.

planes is generally considered to be about 1000 GPa,[55] and with all fibres, quite independent of their origin, a well established correlation exists between preferred orientation and Young's modulus. This is simply that, the more highly oriented the layer planes in the carbon fibre, the higher the tensile modulus.[56] Thus, high modulus with PAN and isotropic pitch precursors may be achieved by heating fibres under tension at temperatures around 2500°C with some stretching for optimum results. As-spun mesophase pitch based carbon fibres are already highly oriented and are capable of being thermoset, carbonized, and graphitized, to achieve moduli, as in the P-120 material (820 GPa), which approach the theoretical limit.[42]

The theoretical tensile strength of a solid is more difficult to evaluate than the theoretical tensile modulus.[57,58] One approach uses the Orowan–Polanyi expression $\sigma_T = (E\gamma_a/a)^{1/2}$ which relates the theoretical strength $\sigma_T$ to the measured Young's modulus ($E$), the surface energy ($\gamma_a$), and the interplanar spacing ($a$) of the planes perpendicular to the tensile axis. The ratio $\sigma/E$ of the measured strength $\sigma$ to the measured modulus varies considerably, but for many materials is in the range 0·1–0·2; even in the most perfect form of carbon, the graphite whisker, the $\sigma/E$ ratio is about 0·03, with $E$ 680 GPa and $\sigma$ about 20 GPa.[58] The strongest carbon fibres have $\sigma/E$ ratios around 0·02. In brittle solids the defects are small cracks; these act as stress concentrators which can grow under the action of an applied stress if they are greater than a critical size determined by the Griffith relationship $\sigma^2 = 2E\gamma_a/\pi C$. If this is applied to a carbon fibre with $\sigma$ 3·9 GPa and $E$ 230 GPa, $C$ is 40 nm, and to a P-55 fibre with $\sigma$ 2·1 GPa and $E$ 380 GPa, $C$ is 230 nm.

The tensile strength and tensile modulus of a material should theoretically rise in proportion; the anomalous decrease in strength noted with these early mesophase pitch fibres exactly parallels similar changes with PAN-based carbon fibres (see Table 1).

### 3.3.1. Effect of Gross Defects

In the field of PAN-based carbon fibres Moreton[59] showed that tensile strength measurements are gauge-length dependent, and that the early fibres had a random distribution of flaws. Chwastiak et al.[49] investigated two samples of carbon fibre prepared from mesophase pitch. The fibres from one sample were apparently free from gross flaws; the fibres from the other contained internal voids. The tensile strength data at different gauge lengths were analysed by Weibull statistics which are based on the 'weakest link' theory of tensile strength. Extrapolations suggested that both samples would have a tensile strength of 7 GPa at 0·1 mm gauge length. This can be considered as an evaluation of the intrinsic strength of the material. In practical terms, the strength of the fibre at 0·3 mm is more important, since this is a good estimate of the load transfer length in a composite material reinforced with carbon fibre. At 0·3 mm the intrinsic strength in the flaw-free fibres reported above can be determined as about 5 GPa. Later, a very detailed Weibull analysis was carried out on mesophase pitch based carbon fibres by Beetz.[60] The predicted intrinsic strength at 0·3 mm gauge length was 3·5 GPa.

Johnson[61] and Johnson and Thorne[62] investigated the fracture surfaces of PAN-based carbon fibres by scanning electron microscopy. Tensile fracture was seen to occur at recognizable flaws located either at the surface or internally; they concluded that these flaws came from impurities originating in the PAN precursor. Sharp and Burnay[63] used a high-voltage TEM to investigate flaws and found that there is no well defined relationship between defect size and fracture strength; indeed, where defects occur in groups, fracture does not always occur at the largest defect. This result immediately raises doubt concerning predictions of tensile strength based on 'weakest link' theories, a doubt which is intensified by the knowledge that PAN-based carbon fibres are available commercially with a tensile strength of 4·5 GPa (see Table 1).

A series of experiments by Moreton and Watt[64] showed that, when rigorous methods of dope filtration and clean-room spinning were applied in the production of PAN precursor fibres, carbon fibres could be produced which showed no decrease in tensile strength as heat treatment temperature and modulus increased. When samples were deliberately contaminated at the final stage of the spinning process, with carbon black, silica and ferric oxide, the individual fibre strengths were lower and the coefficients of variation were greater than for the clean samples. A similar analysis, by Jones et al.,[65] of flaws in high-strength carbon fibres from a contaminated mesophase pitch precursor, used an SEM with an energy dispersive X-ray analyser attached. Fracture surfaces were examined after tensile failure, and were always found to contain recognizable defects. The regions around these defects exhibited traces of contaminants such as iron, nickel and chromium. Fibres prepared under clean conditions failed at higher strengths, for example 3·0–4·0 GPa, as compared to 1·5–2·5 GPa for the contaminated specimens; in addition, it was usually difficult to detect defects in the fracture faces. In the absence of gross defects, surface pits were seen as a strength-limiting feature which would give an intrinsic tensile strength of 3·8 GPa at 0·3 mm gauge length, a prediction very similar to that obtained by Beetz.[60]

It is now widely recognized that defects arising from contaminating particles are the major cause of reduced strength at higher heat-treatment temperatures, and that to evaluate the intrinsic strength of carbon fibres from any source requires a complete understanding of their macro- and micro-structural organization and the behaviour of these structures under tensile deformation.

### 3.3.2. Tensile Failure

There have been many unsuccessful attempts to describe the fracture mechanisms of PAN-based carbon fibres, particularly in terms of dislocation pile-up at grain boundaries, the unbending of curved ribbons, the presence of density fluctuations, or yield processes involving local shear deformation and slippage; these are thoroughly discussed in a review by Reynolds.[66] Although a wide range of internal defects has been observed, no simple relationship could be found between flaw diameter, fibre strength, and surface free energy. This led to the proposal by Reynolds and Sharp[67] of a crystallite shear limit for fibre fracture.

The Reynolds and Sharp mechanism of fracture is based on the idea that crystallites are weakest in shear on the basal planes. When tensile stress is applied to misoriented crystallites locked into the fibre structure, the shear stress cannot be relieved by cracking or yielding between basal planes. The shear strain energy may be sufficient to produce basal plane rupture in the misoriented crystallite, and hence a crack which will

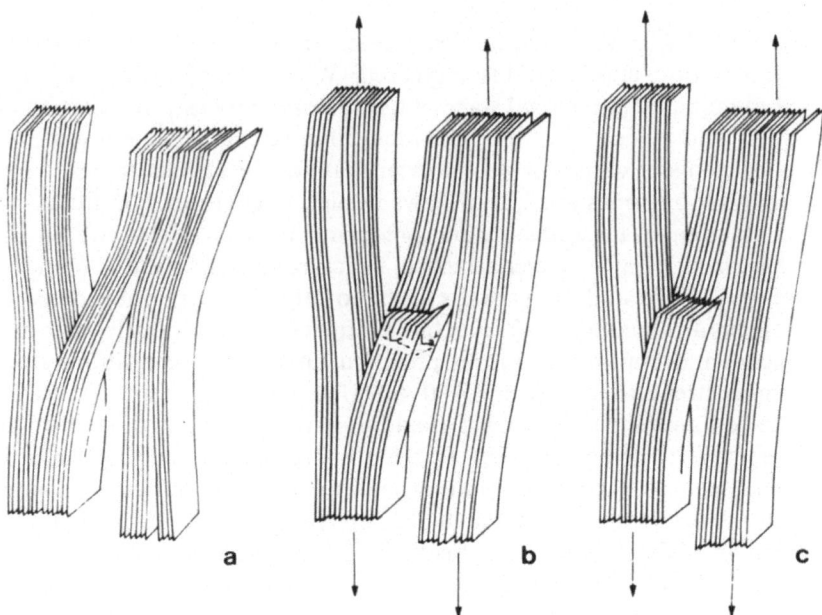

FIG. 13.   Reynolds–Sharp mechanism of tensile failure. (a) Misoriented crystallite linking two crystallites. (b) Tensile stress exerted on crystallites causes basal plane rupture. (c) Crack propagates along $L_a\perp$ and $L_c$ leading to catastrophic failure if the crack size is greater than the critical size.

propagate both across the basal plane and, by transfer of shear stress, through adjacent layer planes. A schematic diagram of a misoriented crystallite well locked into the surrounding crystallites is shown in Fig. 13(a). When stress is applied, basal plane rupture takes place, Fig. 13(b), and proceeds throughout the local region, Fig. 13(c). However, before a crack can propagate through a fibre and cause failure, either one of two conditions must be fulfilled.

(1)   The crystallite size in one of the directions of propagation of a crack, that is either $L_c$ or $L_a\perp$, must be greater than the critical flaw size $C$ for failure in tension (we have already noted that $C$ is about 230 nm for a P-55 mesophase pitch based carbon fibre).

(2)   The crystallite which initiates catastrophic failure must be sufficiently continuous with its neighbouring crystallites for the crack to propagate.

The first condition is not normally fulfilled because both $L_c$ and $L_a\perp$ are much less than $C$, as in Fig. 13, although the effective value of $L_a\perp$ is considerably greater than the values measured by X-ray diffraction where layer planes are curved or 'hairpin' shaped. The second condition is most likely to be satisfied in those regions of enhanced crystallization and misorientation which have been observed around a defect.[63,68] The fortuitous continuity of large crystallites at a large angle of misorientation may well explain why a fibre can fail at the smaller of two defects. Essentially, it is the presence of large misoriented crystallites which cause failure, not the presence of a hole. Alternatively, the second condition of failure is likely to be satisfied where crystallization occurs in sheets.

In a recent study of tensile failure,[69] specimens from an old batch of type I PAN-based carbon fibres, known to contain many flaws, were stressed to failure in glycerol. This enabled the fracture ends to be preserved intact for subsequent examination, first by SEM and then, after embedding and sectioning, by TEM. Internal flaws which did not initiate failure were seen to have walls containing crystallites arranged mainly parallel to the fibre axis. Internal and surface flaws which did initiate failure often showed evidence of large misoriented crystallites in the walls of the flaws. Continuity of crystallites in the walls may well give rise to cracks which exceed the critical flaw size.

Further proof for the concept that large misoriented crystallites, together with continuity of structure in the walls of flaws, cause fibre failure under stress is found in earlier studies of both isotropic pitch-based and lignin-based carbon fibres.[70,71] These fibres, which had very inferior tensile strengths, contained many flaws in the form of inclusions caused by catalytic graphitization around impurity particles in the precursor material. A typical lattice fringe image of one such inclusion is shown in Fig. 14. These inclusions are solids of revolution whose walls are often contiguous with the normal layer-plane structure of the fibre. When the fibre is stressed, there must be misoriented crystallites at various angles which are under tension; consequently, layer-plane failure takes place by the Reynolds–Sharp mechanism, and a crack is initiated which is able to propagate around the wall of the inclusion, thus precipitating fibre failure.

Further evidence in support of the Reynolds–Sharp mechanism can be obtained from studies of acoustic emission carried out during the fatigue testing of carbon fibre reinforced composites.[72] The considerable acoustic energy produced during the first few cycles can be considered to be a result of cracks being produced in the fibre which are below the critical size for failure.

In mesophase pitch based carbon fibres, the high preferred orientation

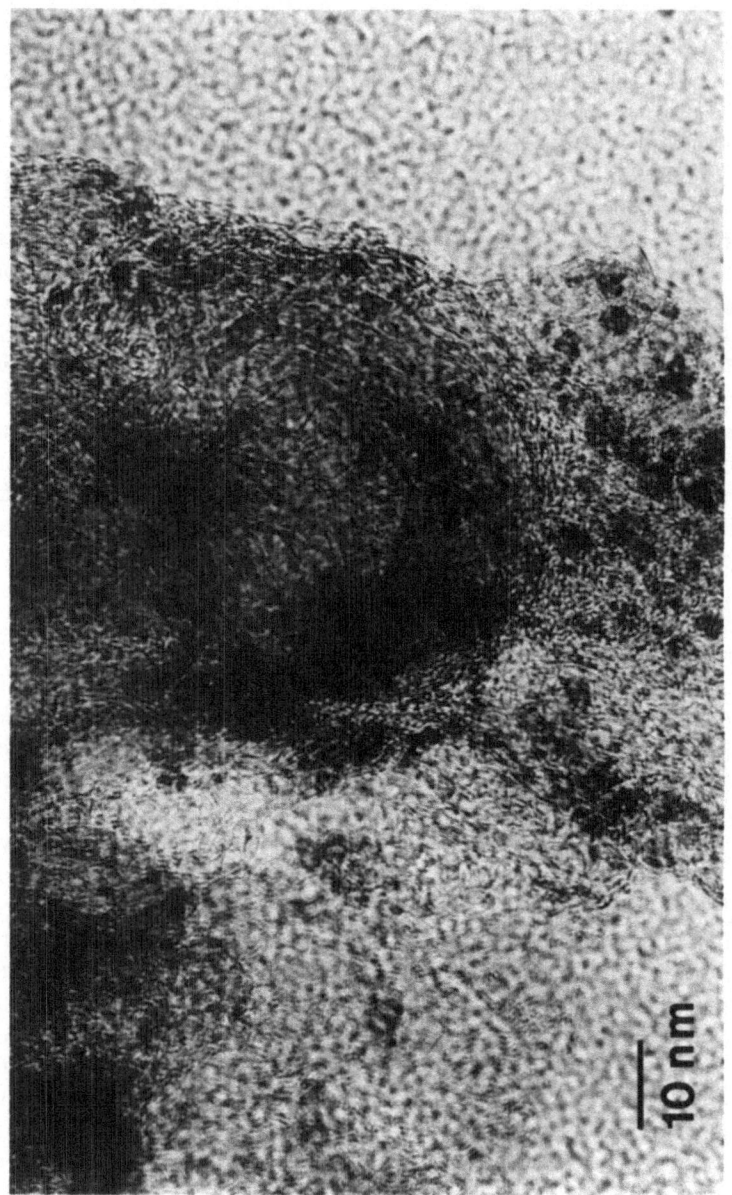

FIG. 14. Lattice fringe image of inclusion in isotropic pitch based carbon fibre.

reduces the number of large misoriented crystallites, so that, in the absence of large flaws with graphitic walls due to catalytic effects, a greater intrinsic strength might be expected than in a similar PAN-based carbon fibre. However, the continuity available to a crack, once initiated, by virtue of the presence of sheet-like organization will lead to a comparatively lower tensile strength.

### 3.3.3. Flexural Failure

Townsend[73] has reported that, when made up in unidirectionally oriented composites, mesophase pitch based carbon fibre does not perform as well as PAN-based carbon fibre when the materials are subjected to compressive or flexural stress. For example, a composite of P-55 fibre has a compressive strength of 0·48 GPa and a flexural strength of 0·83 GPa compared with 1·59 and 1·79 GPa respectively for a composite of PAN-based fibre. In studies of the flexural behaviour of single fibres, DaSilva and Johnson[74] found that mesophase pitch based carbon fibres with $E$ 380 GPa and $\sigma$ 1·90 GPa have a knot strength of 10 MPa compared to 63 MPa for a similar PAN-based carbon fibre, and fail at a diameter of 840 $\mu$m as compared to 370 $\mu$m for the PAN-based carbon fibre. Typical fracture faces after both tensile and flexural failure are illustrated in Fig. 15 for the mesophase pitch based carbon fibre and in Fig. 16 for the PAN-based carbon fibre. The fracture face after flexural failure of the PAN-based fibre is distinctly different from the face after tensile failure; a corrugated area corresponding to a region of maximum shear stress can be seen after flexure. The fracture face of the mesophase pitch based fibre after flexural failure is remarkably similar to the fracture face after tensile failure, revealing sheet-like structures and no evidence for a build-up of shear stress. Evidently the sheet-like structure present in the mesophase pitch based carbon fibre can propagate a transverse crack initiated by the Reynolds–Sharp mechanism far better than the random structure of the PAN-based carbon fibre. When mesophase pitch based carbon fibres are deformed in a knot or under any compressive or flexural loading, a crack is initiated relatively easily and propagated beyond the critical size by the sheet-like crystalline structure.

Tensile and flexural failure in P-55, P-75 and P-100 fibres has also been reported by White and his colleagues.[53] The oriented-core structures seen in P-75 and P-100 fibres are very similar to the structures depicted in Fig. 15. Although loop tests showed some plastic deformation, no evidence was found for a kink mechanism of the type seen in Kevlar. Nevertheless, Jones and Johnson[75] had earlier suggested that a kink mechanism

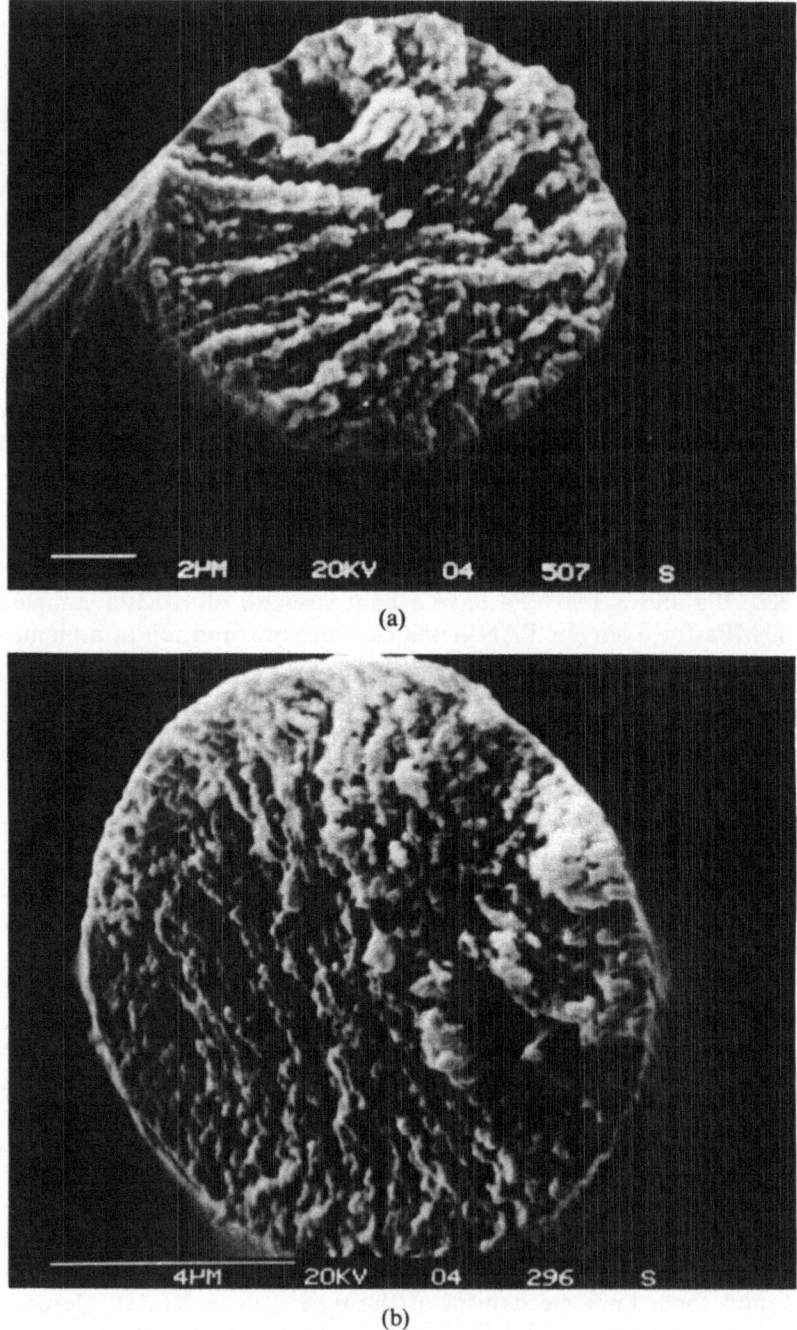

(a)

(b)

FIG. 15. Scanning electron micrographs of fracture faces in mesophase pitch based carbon fibres after (a) tensile failure, (b) flexural failure.

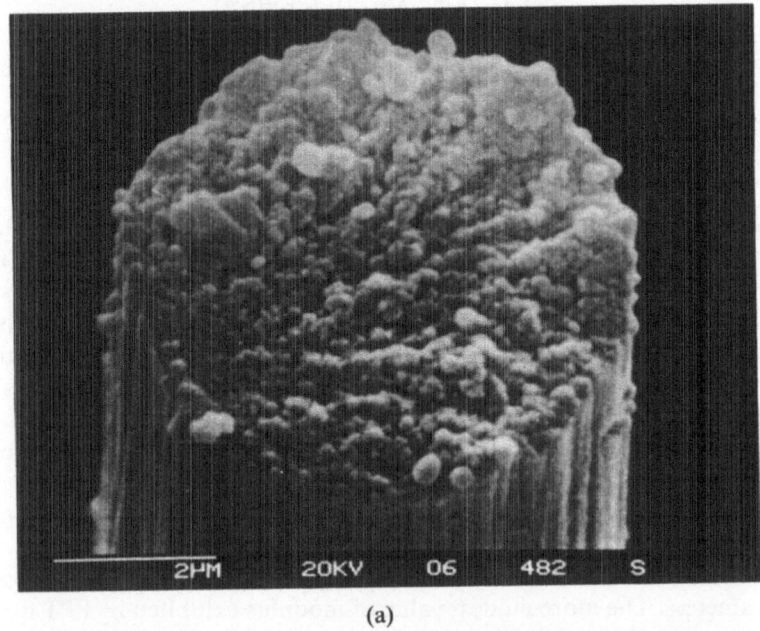

(a)

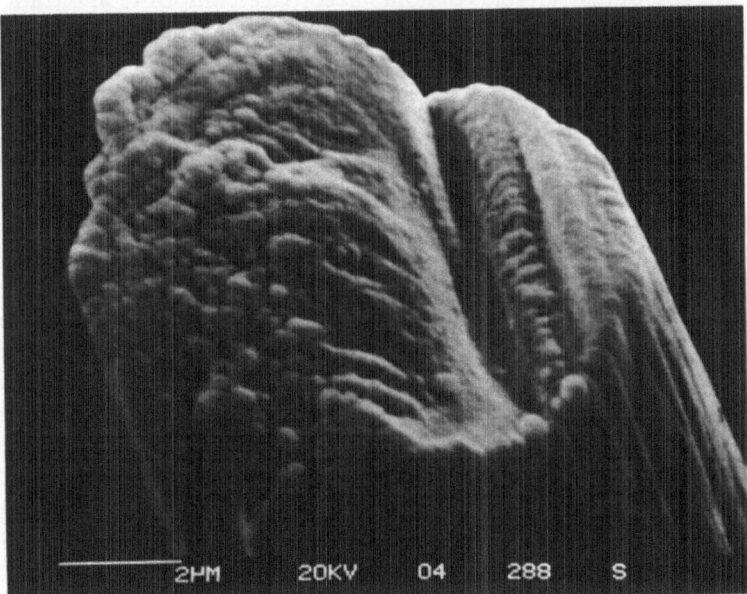

FIG. 16.  Scanning electron micrographs of fracture faces in PAN-based carbon fibres after (a) tensile failure, (b) flexural failure.

operates for PAN-based carbon fibre, although no evidence for such a mechanism could be found in studies by DaSilva and Johnson.[74]

## 4. CONCLUSION

High axial orientation, and thus high tensile modulus, can be obtained in both PPT fibres and pitch-based carbon fibres without the need for a drawing stage, when fibres are spun from liquid-crystal mesophase. In the case of PPT, spinning takes place from a lyotropic system, and in the case of carbon from a thermotropic system. Unlike conventional thermoplastic fibres and PAN-based carbon fibre, both PPT fibres and mesophase pitch based carbon fibres exhibit preferred lateral orientation into sheet-like structures; this leads to relatively low ratios of compressive to tensile strength.

The outstanding characteristic of both PAN-based and mesophase pitch based carbon fibres is their exceptionally high tensile modulus which can be varied considerably according to the choice of processing parameters. The more modest value of modulus exhibited by PPT fibres is offset to some degree by their lower specific gravity, thus providing a high specific modulus suitable for some aerospace composite applications.

Comparison of PPT and carbon fibres also reveals significant differences in fracture behaviour; indeed, the former tend to fibrillate readily whereas the latter are very brittle. This may be attributed to the very different nature of the molecular aggregates; in carbon fibres the layers are formed by strong covalent bonds, whereas in PPT the sheets are stabilized by relatively weak hydrogen bonds. In mesophase pitch based carbon fibres the sheet-like aggregations allow easy crack propagation and on bending the fibres fail at very low strain by a Reynolds–Sharp mechanism. PPT fibres such as Kevlar survive bending intact and, although some fibrillation may occur on the tensile side of a loop or knot, the fibre yields by molecular buckling on the compressive side. Consequently, PPT fibres are of particular value in woven assemblies or components subject to flexure.

Although some mesophase pitch based carbon fibres may have moduli much higher than the PAN-based carbon fibres, they are generally not as useful for composite fabrication owing to their comparatively poor' flexibility. Nevertheless, where high thermal or electrical conductivity is required, mesophase pitch based carbon fibres have an advantage. It is clear therefore that the crystallization properties of liquid-crystal pitch are at present a limiting factor in the usefulness of a most interesting material.

## REFERENCES

1. KWOLEK, S. L. (Du Pont) (1966). Brit. Patent 1,198,081 (priority 13 June, USA).
2. MAGAT, E. E. (1980). *Phil Trans. Roy. Soc.*, **A294**, 463.
3. KWOLEK, S. L. (Du Pont) (1968). Brit. Patent 1,283,064 (priority 12 June, USA).
4. AKZO, N. V. (1975). Brit. Patent 1,547,802 (priority 21 February, Netherlands).
5. ERDEMIR, A. B., JOHNSON, D. J. and TOMKA, J. G. (1986). *Polymer*, **27**, 441.
6. HARAGUCHI, K., KAJIYAMI, T. and TAKAYANAGI, M. J. (1979). *Appl. Polym. Sci.*, **23**, 915.
7. NORTHOLT, M. G. (1974). *Eur. Polym. J.*, **10**, 799.
8. DOBB, M. G., JOHNSON, D. J. and SAVILLE, B. P. (1977). *J. Polym. Sci. Polym. Symp.*, **58**, 237.
9. HERGLOTZ, H. K. (1980). *J. Colloid Interface Sci.*, **75**, 105.
10. DOBB, M. G., JOHNSON, D. J. and SAVILLE, B. P. (1977). *J. Polym. Sci. Polym. Phys. Ed.*, **15**, 2201.
11. HAGEGE, R., JARRIN, M. and SOTTON, M. J. (1979). *J. Microscopy*, **115**, 65.
12. DONALD, A. M. and WINDLE, A. H. (1983). *Colloid Polym. Sci.*, **261**, 793.
13. HORIO, M., KANEDA, T., ISHIKAWA, S. and SHINAMURA, K. (1984). *Sen-i Gakkaishi*, **40**, T285.
14. DOBB, M. G. and ROBSON, R. M. (1985). Private communication.
15. DOBB, M. G., JOHNSON, D. J. and SAVILLE, B. P. (1979). *Polymer*, **20**, 1284.
16. PANAR, M., AVAKIAN, P., BLUME, R. C., GARDNER, K. H., GIERKE, T. D. and YANG, H. H. (1983). *J. Polym. Sci. Polym. Phys. Ed.*, **21**, 1955.
17. MORGAN, R. J., PRUNEDA, C. O. and STEELE, W. J. (1983). *J. Polym. Sci. Polym. Phys. Ed.*, **21**, 1757.
18. CAPACCIO, G., GIBSON, A. G. and WARD, I. M. (1979). In: *Ultra-high modulus polymers* (Ed. A. Ciferri and I. M. Ward), Applied Science Publishers, Ch. 1.
19. CARTER, G. B. and SCHENK, V. T. (1975). *Structure and properties of oriented polymers*, Applied Science Publishers.
20. BLACK, W. B. and PRESTON, J. (1973). *High modulus wholly aromatic fibres*, Marcel Dekker, New York.
21. NORTHOLT, M. G. and VAN AARTSEN, J. J. (1977). *J. Polym. Sci. Polym. Symp.*, **58**, 283.
22. NORTHOLT, M. G. and HOUT, R. V. D. (1985). *Polymer*, **26**, 310.
23. BUNSELL, A. R. (1975). *J. Mater. Sci.*, **10**, 1300.
24. COOK, J., HOWARD, A., PARRATT, N. J. and POTTER, K. D. (1982). *Proc. 3rd Ann. Riso Symp. Metallurgy and Mater. Sci.*, Denmark, p. 193.
25. DOBB, M. G. and MCINTYRE, J. E. (1984). *Adv. Polym. Sci.*, **60/61**, 63.
26. GREENWOOD, J. H. and ROSS, P. G. (1974). *J. Mater. Sci.*, **9**, 1809.
27. DOBB, M. G., JOHNSON, D. J. and SAVILLE, B. P. (1981). *Polymer*, **22**, 960.
28. DETERESA, S. J., FARRIS, R. J. and PORTER, R. S. (1982). *Polymer Composites*, **3**, 57.
29. SCHOPEE, N. M. and SKELTON, J. (1974). *Text. Res. J.*, **44**, 968.
30. PHOENIX, S. L. and SKELTON, J. (1974). *Text. Res. J.*, **44**, 934.
31. SKELTON, J. (1975). *Proc. 5th Int. Wool Textile Res. Conf.*, Aachen, **2**, 362.
32. VOLK, H. F. (1976). *Proc. 4th London Int. Conf. Carbon and Graphite, 1974*, Society of Chemical Industry, London, p. 183.
33. OTANI, S. (1965). *Carbon*, **3**, 3.

34. OTANI, S. (1965). *Carbon*, **3**, 213.
35. OTANI, S., YAMADA, K., KOITABASHI, T. and YOKOYAMA, A. (1966). *Carbon*, **4**, 425.
36. OTANI, S., YOKOYAMA, A. and NUKUI, A. (1969). *Appl. Polym. Symp.*, **9**, 325.
37. ARAKI, T. and GOMI, S. (1969). *Appl. Polym. Symp.*, **9**, 331.
38. HAWTHORNE, H. M., BAKER, C., BENTALL, R. H. and LINGER, K. R. (1970). *Nature*, **227**, 946.
39. BROOKS, J. D. and TAYLOR, G. H. (1965). *Carbon*, **3**, 185.
40. BROOKS, J. D. and TAYLOR, G. H. (1965). *Nature*, **206**, 697.
41. BARR, J. B., CHWASTIAK, S., DIDCHENKO, R., LEWIS, I. C., LEWIS, R. T. and SINGER, L. S. (1976). *Appl. Polym. Symp.*, **29**, 161.
42. SINGER, L. S. (1978). *Carbon*, **16**, 409.
43. US PATENT 3,976,729 (1976); 4,026,788 (1977).
44. DIEFENDORF, R. J. (1984). IN: *Polymers for fibres and elastomers* (Ed. J. C. Arthur Jr), ACS Symposium Series No. 260, American Chemical Society, Washington, p. 209.
45. RIGGS, D. M. (1984). Reference 44, p. 245.
46. JOHNSON, D. J. and TYSON, C. N. (1970). *J. Phys. D, Appl. Phys.*, **3**, 526.
47. TAYLOR, G. A. (1985). Private communication.
48. RULAND, W. and PLAETSCHKE, R. (1985). *Proc. 17th Biennial Conf. Carbon*, Lexington, American Carbon Soc. and University of Lexington, p. 356.
49. CHWASTIAK, S., BARR, J. B. and DIDCHENKO, R. (1979). *Carbon*, **17**, 49.
50. WHITE, J. L., NG, C. B., BUECHLER, M. and WATTS, E. J. (1981). *Proc. 15th Biennial Conf. Carbon*, Philadelphia, American Carbon Soc. and University of Pennsylvania, p. 310.
51. HARRIS, W. F. (1977). *Sci. Amer.*, **237**, 130.
52. ZIMMER, J. E. and WHITE, J. L. (1983). *Adv. Liq. Cryst.*, **5**, 157.
53. NG, C. B., HENDERSON, G. W., BUECHLER, M. and WHITE, J. L. (1983). *Proc. 16th Biennial Conf. Carbon*, San Diego, American Carbon Soc. and University of California, p. 515.
54. HAWTHORNE, H. M. (1971). *Proc. 1st Int. Conf. Carbon Fibres*, Plastics Institute, London, p. 81.
55. DOREY, G. (1980). *Phys. Technol.*, **11**, 56.
56. FOURDEUX, A., PERRET, R. and RULAND, W. (1971). *Proc. 1st Int. Conf. Carbon Fibres*, Plastics Institute, London, p. 57.
57. MACMILLAN, N. H. (1972). *J. Mater. Sci.*, **7**, 239.
58. BACON, R. (1960). *J. Appl. Phys.*, **31**, 283.
59. MORETON, R. (1968). *Fibre Sci. Technol.*, **1**, 273.
60. BEETZ, C. P. (1982). *Fibre Sci. Technol.*, **16**, 45, 81.
61. JOHNSON, J. W. (1969). *Appl. Polym. Symp.*, **9**, 229.
62. JOHNSON, J. W. and THORNE, D. J. (1969). *Carbon*, **7**, 659.
63. SHARP, J. V. and BURNAY, S. G. (1971). *Proc. 1st Int. Conf. Carbon Fibres*, Plastics Institute, London, p. 68.
64. MORETON, R. and WATT, W. (1974). *Nature*, **247**, 360.
65. JONES, J. B., BARR, J. B. and SMITH, R. E. (1980). *J. Mater. Sci.*, **15**, 2455.
66. REYNOLDS, W. N. (1973). In: *Chemistry and physics of carbon* (Ed. P. L. Walker and P. A. Thrower), Vol. 11, Dekker, New York, p. 1.
67. REYNOLDS, W. N. and SHARP, J. V. (1979). *Carbon*, **12**, 103.

68. BENNETT, S. C. and JOHNSON, D. J. (1979). *Carbon*, **17**, 25.
69. BENNETT, S. C., JOHNSON, D. J. and JOHNSON, W. (1983). *J. Mater. Sci.*, **18**, 3337.
70. JOHNSON, D. J., TOMIZUKA, I. and WATANABE, O. (1975). *Carbon*, **13**, 529.
71. JOHNSON, D. J., TOMIZUKA, I. and WATANABE, O. (1975). *Carbon*, **13**, 321.
72. BUNSELL, A. R., FUWA, M. and HARRIS, B. (1975). *J. Phys. D*, **8**, 1460.
73. TOWNSEND, H. N. (1980). *Proc. 3rd Int. Conf. Compos. Mater.*, **1**, 453.
74. DASILVA, J. L. G. and JOHNSON, D. J. (1984). *J. Mater. Sci.*, **19**, 3201.
75. JONES, W. R. and JOHNSON, J. W. (1971). *Carbon*, **9**, 645.

*Chapter 5*

# STRUCTURE AND PROPERTIES OF THERMOTROPIC LIQUID CRYSTALLINE COPOLYESTERS

J. BLACKWELL and A. BISWAS

*Department of Macromolecular Science,
Case Western Reserve University,
Cleveland, Ohio, USA*

## 1. INTRODUCTION

The phenomenon of liquid crystallinity was discovered by Reinitzer[1] in 1888 during a study of the melting behavior of cholesteryl esters, and the structure–property relationships in low molecular weight liquid crystalline (LC) systems are now well understood (see e.g. References 2–4). Liquid crystallinity in higher molecular weight organic substances was reported more recently: in the 1950s, LC behavior was observed in solutions of biopolymers, such as poly(γ-benzyl-L-glutamate), hydroxypropyl-cellulose and tobacco mosaic virus.[5] The role of liquid crystallinity in Nature has evoked a lot of interest because of its contribution to the spatial organization of macromolecules in the living organism, e.g. in the packaging of DNA in chromosomes and the aggregation of microtubules to form the structural framework of the cell.[6]

Since these first studies of biological macromolecules, there has been growing interest in synthetic polymers that show LC order, especially with regard to the production of high performance materials. In the 1960s it was reported, notably by Kwolek[7] of DuPont, that certain wholly aromatic polyamides such as poly(*para*-phenylene terephthalamide) exhibited anisotropic properties in solution, and this led ultimately to the development of the aramid fiber Kevlar. In 1972, Economy and

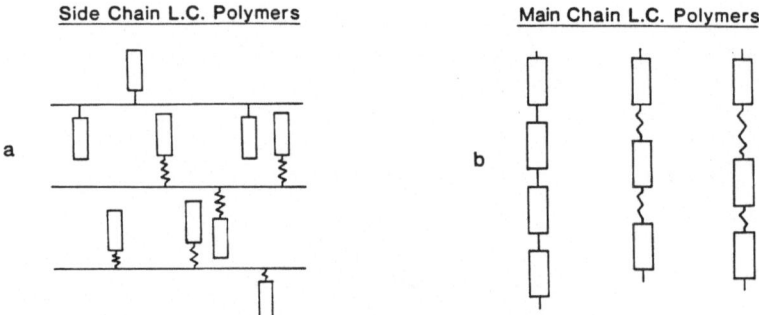

FIG. 1. The structure of polymeric liquid crystals: (a) with mesogens in side chains connected directly or by spacers; (b) with mesogens in the backbone connected directly or by spacers.

coworkers[8] at Carborundum patented a melt-processable aromatic copolyester, prepared from *p*-hydroxybenzoic acid (HBA), 4,4'-biphenol and terephthalic acid (TPA), which is now marketed as Xydar by Dartco. Two years later, an aromatic–aliphatic copolyester based on HBA and poly(ethylene terephthalate) (PET) was reported by Kuhfuss and Jackson[9] of Eastman Kodak. Since then, there has been considerable research on liquid crystallinity in polymeric systems. Academic interest has focussed largely on characterizing these systems and understanding their structure–property relationships.[10–13]

For a polymer to exhibit liquid crystalline character, it is necessary for it to have a degree of conformational rigidity. This can be achieved by polymerization of rod-like or plate-like monomers (mesogens), but can also result when the polymer adopts a rigid (helical) conformation. In the so-called side-chain LC polymer (Fig. 1(a)), the mesogens are attached to a flexible polymer backbone.[14–17] The various reported nematic, smectic and cholesteric structures occur as a result of stacking of the side-chains.[18] The alternative is to incorporate the mesogen in the backbone to form a main-chain LC polymer (Fig. 1(b)). The thermotropic liquid crystalline polyesters (LCPE) that are the subject of this review fall into the latter class.

Theoretical studies by Flory[19] and DeMarzio[20] predict that incorporation of rigid or semi-rigid units into the chain backbone will lead to the formation of anisotropic phases in solution or the melt, depending on the conditions. The requirements for the structure of the mesogenic units are relatively broad. LC phases can be obtained by linking prolate stiff units which do not themselves form liquid crystalline structures before polymerization.[18] As expected, the reorientational motions of the main chain are very slow or even totally restricted when the mesogens are linked end-to-end. In many LC polymers, the mesogens are linked by flexible spacer groups, usually $[-CH_2-]_n$ units of constant or variable length. The lengths of the rigid and flexible segments serve to determine the structures and the transition temperatures of the resulting LC phases. The stability of the nematic phase increases with increasing length to diameter ($L/D$) ratios of the rigid unit. The shortest reported rigid rod length giving rise to nematic behavior is 1·1 nm.[21] Increasing spacer lengths make the smectic phase more stable than the nematic phase.

Probably the best known example of a main-chain LC polymer is the wholly aromatic polyamide poly($p$-phenylene terephthalamide)

$$\left[ HN-\!\!\left\langle\bigcirc\right\rangle\!\!-NH-CO-\!\!\left\langle\bigcirc\right\rangle\!\!-CO \right]_n$$

which is marketed by DuPont as Kevlar 49. The absence of flexible spacers and the extended chain conformation, which is inevitable in view of the 1,4 phenylene linkages and the planar amide groups, lead to a nematic mesophase in solution. The polymer is processed as high strength, high modulus fibers by dry-jet wet spinning from concentrated sulfuric acid solutions. A lot of attention is now being given to LC aromatic polyesters, which have analogous structures, but are melt processable. These systems usually exhibit nematic phases in the melt, from which high strength fibers and extrudates can be obtained. The following sections describe the synthesis, properties and characterization of main-chain LC polyesters.

## 2.  SYNTHESIS OF LC POLYESTERS

The usual synthetic routes for the production of aliphatic polyesters can also be utilized to obtain LC aromatic polyesters. Research in mesophase polyesters was pioneered in 1965–75 by Goodman (ICI), Economy (Carborundum), Schaefgen (DuPont), Jackson (Tennessee Eastman) and Calundann (Celanese). Some of the first thermotropic polyesters were prepared from diphenols and dicarboxylic acid chlorides by interfacial polymerization or by high temperature solution polymerization. Later it became usual to use an ester exchange reaction between acetoxyaryl groups and carboxylic acid groups, with the elimination of acetic acid at about 200–350°C (melt acidolysis).[22] The bulk of the reaction and acetic acid evolution is accomplished at normal atmospheric pressures, after which the resulting oligomers are polymerized to higher molecular weights by application of a vacuum. Catalyzed melt polymerization via diphenyl esters and the evolution of phenol has also been used. Acidolysis in inert heat exchange media is used in the case of very high melting polymers. Direct esterification of diacids and diols in the presence of a suitable catalyst has also been reported.[23] Another approach has been to use transesterification reactions, starting with the phenyl esters of diacids and the aryl diols or with the acyl esters of the aryl diols. The best-known example of the latter approach is the acidolysis of PET with *p*-acetoxybenzoic acid. The two react together at 275°C to give short acetoxy-terminated and carboxy-terminated chain segments, and these are condensed to a high molecular weight polyester by heating under reduced pressure.[24]

Further details of the synthesis of LCPEs existing in various phases can be found elsewhere.[25] In particular, the synthesis of wholly aromatic LCPEs for high performance fibers and plastics is the subject of numerous patents.[26-29] These systems will be discussed in greater detail below.

### 2.1. Polyesters with Flexible Spacers

Numerous nematic, smectic and cholesteric polyesters in which the mesogens are separated by flexible spacers have been synthesized and characterized. The spacers are usually sequences of the type $-\!\!\left[CH_2\right]_n$ or $-\!\!\left[CH_2CH_2O\right]_n$.[30] Very interesting trends in the structure and properties have been noted for these systems, depending on the length and placement of the rigid and flexible units.[31,32] Flory and Ronca[33] predict the occurrence of a crystal–nematic transition in systems where the axial ratio $(L/D)$ of the rigid unit does not exceed 6·4. However, a nematic phase may also occur for higher axial ratios when a distribution of flexible spacer lengths is used. Van Luyen and Strzelecki[34] used the following mesogens to investigate the effect of altering the size of the rigid block while keeping the spacer length constant:

$\sim 13\,\text{Å}$

$\sim 19\,\text{Å}$

$\sim 25\,\text{Å}$

As the length of the rigid block was increased from $\sim 13\,\text{Å}$ to $\sim 25\,\text{Å}$, while the spacer length was set at 9 methylene units, the nematic–isotropic transition temperature $T_{NI}$ increased from 200°C to 350°C. Another study by Ober et al.[35] showed that the tacticity of the mesogenic group in the chain also determines mesophase formation. Random head–tail orientations of the mesogenic dyad

$$-\text{OCO}-\!\!\bigcirc\!\!-\text{OCO}-\!\!\bigcirc\!\!-\text{COO}-$$

resulted in a nematic phase for the polymer. In contrast, the syndiotactic polymer (with an exactly alternating head–tail arrangement in successive dyads) did not form a mesophase.

An odd–even effect on the crystal–nematic ($T_{CN}$) and nematic–isotropic ($T_{NI}$) transition temperatures has been observed with increasing number of methylene units in the spacer group. As shown in Fig. 2, $T_{CN}$ and $T_{NI}$ for the polyester

$$-\text{CO}-(\text{CH}_2)_n-\text{CO}-\text{O}-\!\!\bigcirc\!\!-\text{C}(\text{CH}_3)\!=\!\text{CH}-\!\!\bigcirc\!\!-\text{O}-$$

both have higher values when $n$ is even than for the neighboring odd $n$ polymers.[36] This observation is thought to be related to the conformation of polymethylene segments.[30] Similar behavior has been seen in low molecular weight liquid crystals,[2] and in the melting temperatures of non-mesomorphic polymers such as polyurethanes.[37]

## 2.2. Wholly Aromatic Polyesters

The fiber tensile modulus of an aliphatic–aromatic polyester can be related to an empirical index expressing the 'degree of aromaticity' of the polymer,[23] which is simply the ratio of the number of $sp^2$ hybridized carbons to the total number of non-hydrogen atoms in the repeat. From Fig. 3 it is evident that wholly aromatic polyester structures should have the highest fiber tensile moduli. Such structures should not only have inherent LC character but should also possess the desired stiffness unobtainable from more flexible structures. When rigid units having chain-continuing bonds that are either collinear (as in $p$-phenylene) or parallel and oppositely directed (e.g. 2,6-naphthylene) are linked by the ester group, which has partial double-bond character and limited rotatability, this leads to a stiff extended chain conformation imparting the

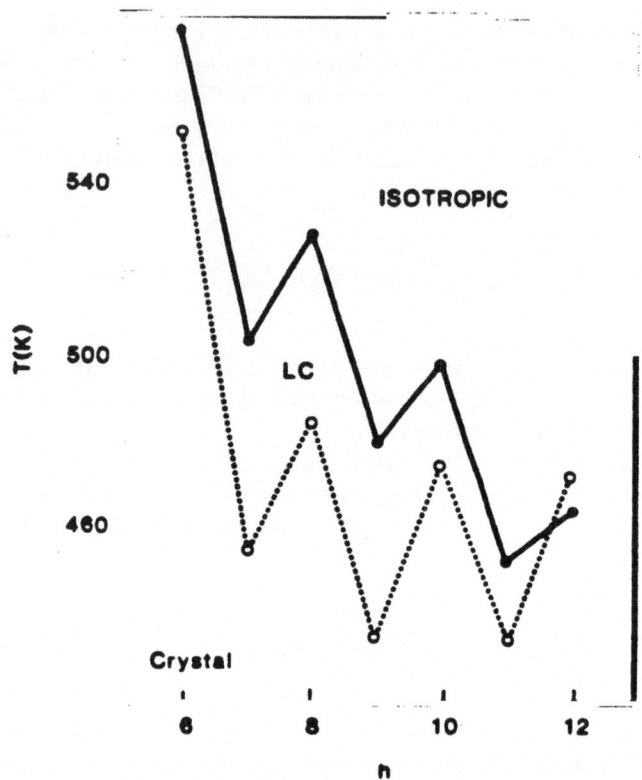

FIG. 2. Variation of $T_{CN}$ and $T_{NI}$ with the number of methylene units for homopolyesters of 4,4'-dihydroxy-$\gamma$-methylstilbene and aliphatic acids.[36]

desired properties to the structure. However, homopolyesters of this type, e.g. poly($p$-hydroxybenzoic acid),

$$+O-\underset{\phantom{}}{\bigcirc}-CO+_{\overline{n}} \qquad \text{poly($p$-HBA)}$$

are infusible, largely intractable crystalline solids. In order to lower the melting points to facilitate melt processing, it is necessary to modify the molecular architecture. The melting point $T_m$ is given by

$$T_m = \Delta H_f / \Delta S_f \tag{1}$$

where $\Delta H_f$ is the enthalpy and $\Delta S_f$ is the entropy of fusion.[38] $\Delta H_f$ is directly related to the degree of crystallinity or order in the system. It is

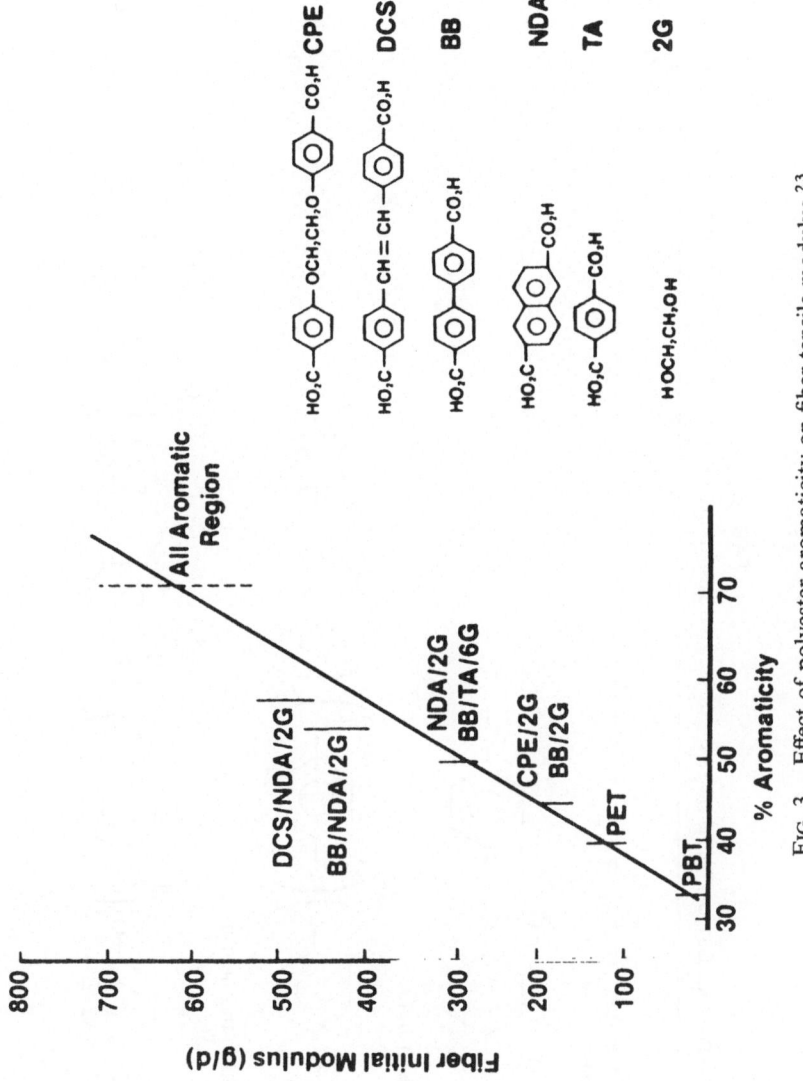

FIG. 3. Effect of polyester aromaticity on fiber tensile modulus.[23]

## TABLE 1

TYPICAL LIST OF THE EFFECTS OF DIFFERENT STRATEGIES TO REDUCE $T_m$ (MODIFIED FROM REFERENCE 48)

| | Melting temp. |
|---|---|
| Copolymer | $> 600°C$ |
| Crank-shaft unit | $\sim 400°C$ |
| Crank-shaft and bent unit | $\sim 350°C$ |

Small substituent  >400°C

Large substituent  ~340°C

Aliphatic unit  ~210°C

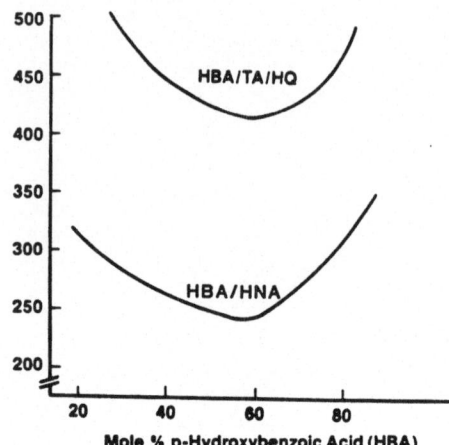

FIG. 4.  Variation of $T_m$ with composition for two copolyesters with and without a crankshaft-like moiety (modified from Reference 23).

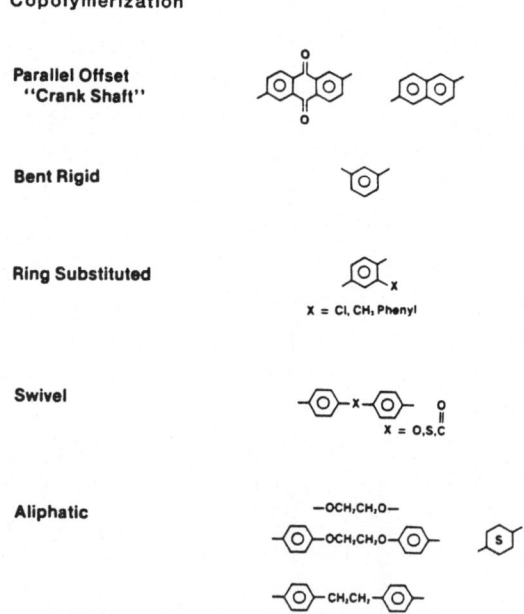

FIG. 5.  Strategies for lowering $T_m$ in LCPEs.[23]

also sensitive to molecular cohesion and chain irregularities. $\Delta S_f$ defines the increase in the disorder in the system during its transition from the solid to the molten (LC) state. In order to reduce $T_m$, one must either lower $\Delta H_f$ and/or increase $\Delta S_f$. In LC systems, $\Delta S_f$ is low due to the inherent order which does not undergo appreciable disruption on melting. Decreasing chain stiffness will lead to higher $\Delta S_f$ values. $\Delta H_f$, on the other hand, can be decreased by introducing defects or irregularities into linear chain segments and by diminishing intra- and inter-chain bonding.

The general methods used for lowering $T_m$ in LCPEs are illustrated in Table 1. The first obvious choice would be to introduce a comonomer. Copolymerization with collinear rigid units to form random copolymers does result in the reduction of $T_m$. However, this reduction may not be sufficient to allow for fiber extrusion, even though there are no interchain hydrogen bonds (as in the polyaramids), and hence further disruption of the main chain order is usually required.[21] This can be achieved by using comonomer units which have parallel but offset chain-continuing bonds, which leads to a substantial decrease of $T_m$. Figure 4 shows plots of $T_m$ vs. molar % composition for two different copolyesters with and without crankshaft-like moieties.[23] These data are typical of those for many LCPEs. Melt processing is easier around the minimum of $T_m$, which occurs at intermediate mole ratios. Incorporation of 'bent' or 'kinked' units imposes an even greater deviation in chain linearity, leading to further reduction in $T_m$. However, introduction of large amounts of such units, with linkage bonds inclined at 60–80°, can lead to a loss of liquid crystallinity.[22] The most commonly used 'bent' units are *meta*- and *ortho*-linked phenylenes, and 1,6 and 2,5 linked naphthylenes.[23,30,39,40] Asymmetrically substituted ring structures are also effective in reducing $T_m$.[22,30,31] Depending on the type of substituent, two opposing effects come into play: steric separation of neighboring chains and dipole–dipole interactions; note that increasing the latter leads to higher values for $T_m$.[41] It has been found that non-linear comonomers are more effective in lowering $T_m$ than substituted *para*-phenyl groups, except in the case of large substituents, e.g. toluyl groups. The use of 'swivel' bonds, such as $-CH_2-$, $-O-$ or $-S-$, and flexible aliphatic spacers increase $\Delta S_f$, thereby reducing $T_m$. The best example for the latter category is HBA-modified PET.[24] However, the melt is not liquid crystalline when the HBA content is less than 30%. Figure 5 summarizes the effects of the different copolymerization strategies on $T_m$. Patents exist for LCPEs corresponding to all of these types of copolymers, and a representative list is given in Table 2.

TABLE 2

LCPEs DEVELOPED BY THE INDUSTRY, INCORPORATING VARIOUS STRATEGIES TO CONTROL $T_m$

| Company | Structural units |
|---|---|
| Du Pont | |
| Du Pont | |
| Du Pont | |
| Eastman Kodak | |

Eastman Kodak

Celanese

Celanese

Celanese

Carborundum

Carborundum

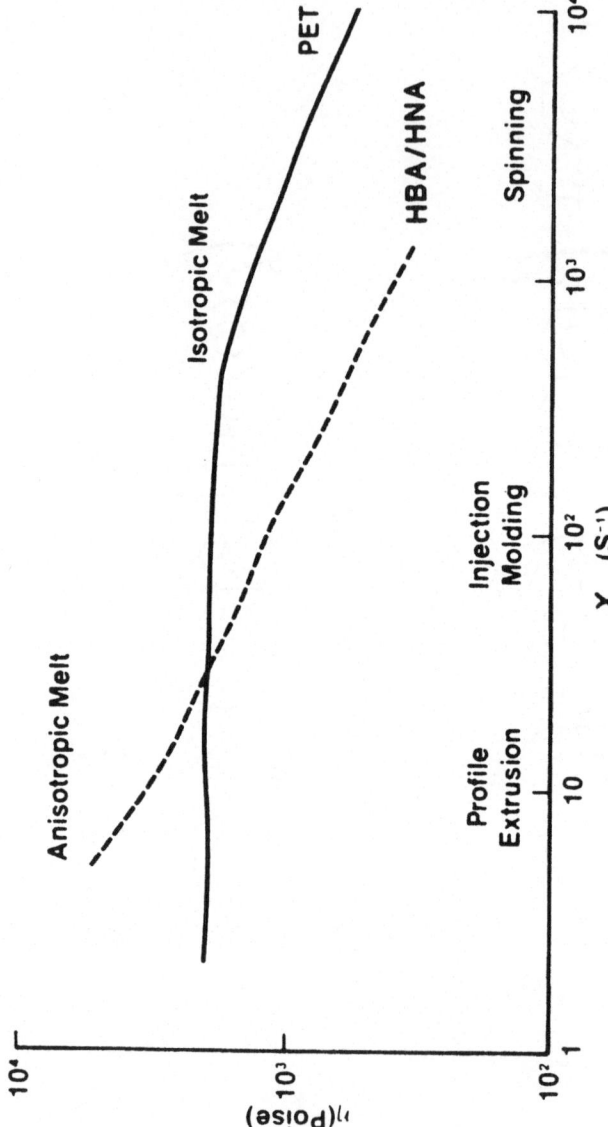

FIG. 6.  Comparison of viscosity–shear rate behavior of conventional and anisotropic melts.[23]

## 3.  PROPERTIES OF LIQUID CRYSTALLINE COPOLYESTERS

### 3.1. Rheology

The 'anomalous' flow phenomena observed for LCPE melts include the occurrence of a yield stress, a large secondary shear-stress maximum after start-up of steady shear, a transient negative normal stress, and shear thickening of the melt.[42] The steady shear viscosity for a polymer in the nematic phase is lower than in the isotropic phase, despite the lower temperatures of measurement for the former case.[21] At the onset of nematic behavior, the melt viscosity of the LC copolyester

is less than that of a similar but non-mesogenic polymer by over three orders of magnitude.[21] Figure 6 compares the viscosity vs. shear rate behavior of PET and the copolymer of HBA and 2,6-dihydroxynaphthalene (HNA). Notice the high shear rate dependence of viscosity for the latter.

Melt viscosity is also composition dependent. Figure 7 shows the variation of melt viscosity of the HBA/PET copolymer at 275°C, at high and low shear rates, as a function of % HBA.[24] It is believed that LC domains[43] are present in the melt and that these domains 'tumble' and remain random. The growth of shear stress observed at the inception of shear flow reflects the orientation of these domains.[44] Better orientation is achieved when shear flow is converted to elongational flow. The resulting extrudates have high tensile moduli and almost negligible linear expansion coefficients.[45]

The melt rheology has also been found to be dependent on the thermal history of the system. Heating the polymer to a high temperature for a brief period, and then returning rapidly to a much lower extrusion temperature, leads to a reduction in the melt viscosity. On the other hand, an increase in viscosity is observed if the sample is held beforehand at a temperature lower than that used for measurement.[46] These changes in viscosity are presumably linked to the variations in LC order arising from

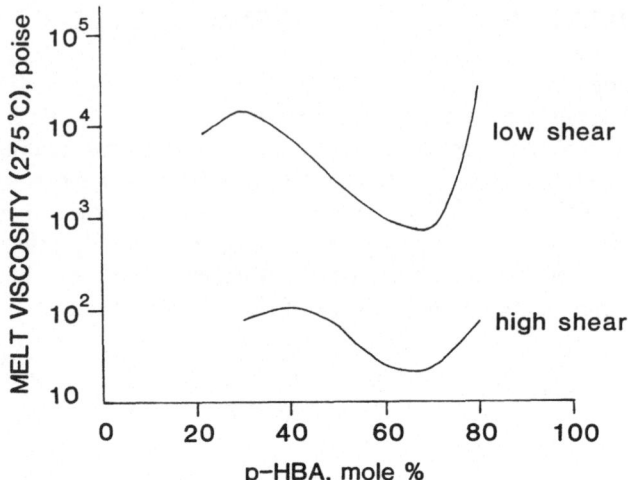

FIG. 7. Effect of composition on melt viscosity at different shears for the HBA/PET copolymer.[24]

differences in the thermal history. LCPE melts are also highly elastic but exhibit negligible die swell after extrusion through capillaries.[21]

## 3.2. Mechanical Properties

The development of LCPEs has been primarily motivated by the expectation that they will yield materials with excellent mechanical properties. The potential uses for these polymers are as melt-spun fibers and molded plastics.

### 3.2.1. Fibers

The properties of fibers prepared from some of the wholly aromatic LCPEs compare well with those of other high strength organic reinforcements, including Kevlar 49. Figure 8 shows the stress–strain curves for several types of fibers. Table 3 lists values of tensile strengths and moduli for various high strength organic fibers, and it can be seen that the wholly aromatic LCPEs have properties comparable to those of Kevlar 49. LCPEs have a density of $\sim 1.4\,\mathrm{g\,cm^{-3}}$, and thus the properties of LCPE fibers on a weight basis are superior to those of glass or steel. As a result of the extended chain conformations, the elongations at break are quite small ($\sim 3$–4%).[21] During fiber spinning, the mechanical properties develop quickly with increasing drawdown ratio (Fig. 9); development of orientation and crystallinity is complete within a small distance from the

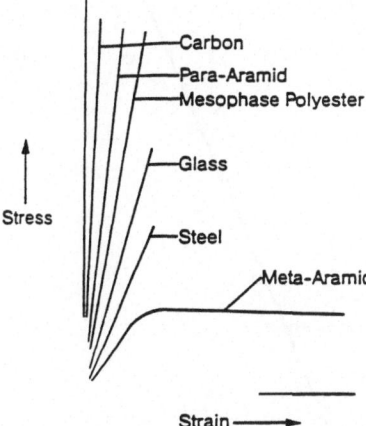

FIG. 8.    Stress–strain curves for several types of fibers.[22]

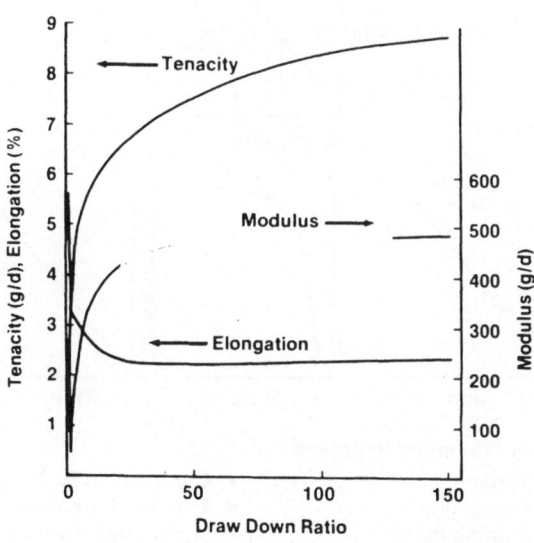

FIG. 9.    Development   of   properties   of   copoly(HBA/HNA)   fibers   with drawdown.[23]

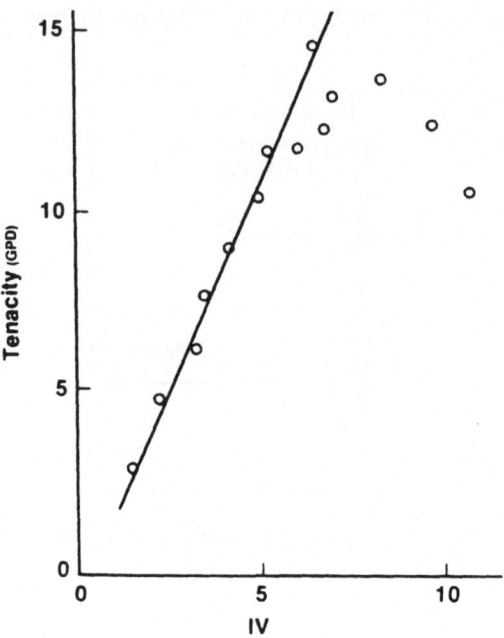

FIG. 10. Relationship between molecular weight (IV) and fiber tensile strength.[23]

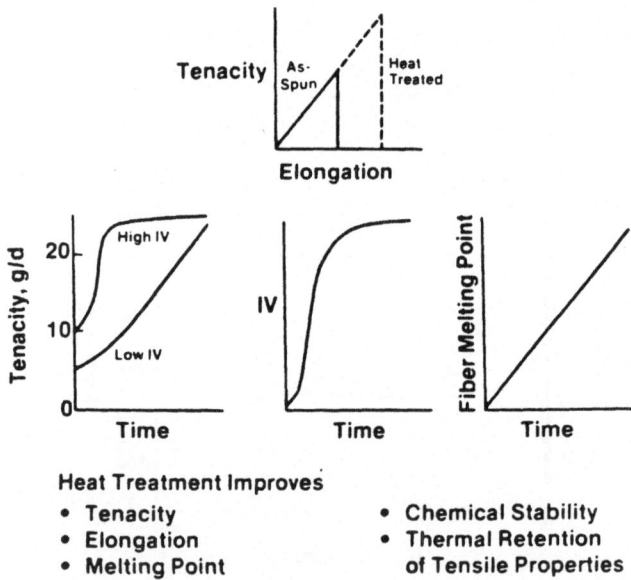

**Heat Treatment Improves**
- Tenacity
- Elongation
- Melting Point
- Chemical Stability
- Thermal Retention of Tensile Properties

FIG. 11. Summary of the property changes observed after high temperature annealing of melt-spun LCPE fibers.[23]

## TABLE 3

TENSILE STRENGTHS AND MODULI FOR VARIOUS HIGH-STRENGTH ORGANIC FIBERS

| Polymer | Strength (GPa) | Modulus (GPa) |
|---|---|---|
| Kevlar 49 | 2·6 | 119 |
| Kevlar 29 | 2·9 | 59·5 |
| PBO | 0·6 | 84·6 |
| PBT | 2·4 | 250 |
| (HBA/PET) (60/40) | 0·4 | 23·3 |
| (HBA/HNA) (75/25) | 2·4 | 65·5 |
| (HBA/DHN/TPA) (60/20/20) | 2·2 | 68·6 |
| (HNA/HQ/TPA) (82/9/9) | 2·7 | 60·7 |

spinnerette. LCPEs having non-linear structural units generally have lower tenacities, moduli and higher elongations at break.

The tensile strengths depend on the molecular weights of the polymers, which are measured as IV in Fig. 10. Molecular weights of $\sim 25\,000$ have been reported in patent literature for as-spun fibers of copoly-(HBA/HNA), based on studies of solutions in pentafluorophenol.[47] As-spun LCPE fibers are not drawable in the conventional sense, but fiber properties can be enhanced by heat treatment or annealing. Typically, annealing is done at zero tension in a flowing dry air or inert gas environment for 1–2 h at temperatures 10–30°C below the melting point of the as-spun fiber.[21] Figure 11 summarizes the changes that occur in the properties as a result of such annealing. (The data in Table 3 are for heat-treated LCPE fibers.) It is believed that the heat treatment results in further solid-state polymerization, as indicated by the higher IVs, and leads to the removal of critical flaws in the system, thereby improving the mechanical properties as well as thermal and chemical resistance. The melting temperature also rises, and there is an increase in three-dimensional order, which is substantiated by the X-ray diffraction patterns.[23] Figure 12 shows that the heat-treated fiber retains its properties over a wide range of temperatures, more so than the as-spun fiber or a conventional PET fiber.

The dynamic mechanical spectrum of the copoly(HBA/HNA) fiber (Fig. 13) reveals three loss processes at $-50$°C, 41°C and 110°C, which are labelled as the $\gamma$, $\beta$ and $\alpha$ transitions respectively. The $\gamma$ transition has been assigned to reorganizational motions of the phenylene groups, and the $\beta$ transition to rotational motion of several monomer units. The $\alpha$ transition

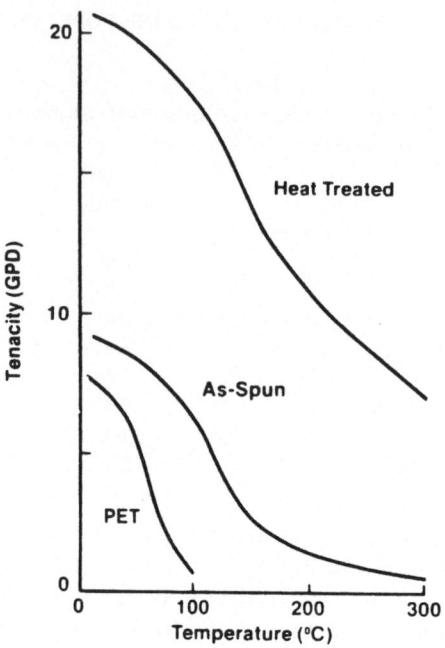

FIG. 12. Variation of fiber tensile strength with temperature for as-spun and heat-treated LCPE fiber compared to a conventional fiber.[24]

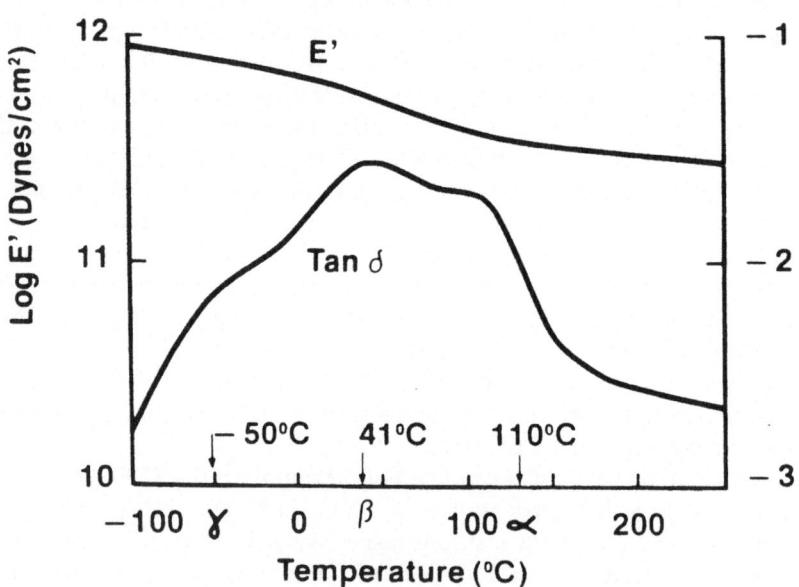

FIG. 13. Dynamical mechanical behavior of LCPE fiber.[23]

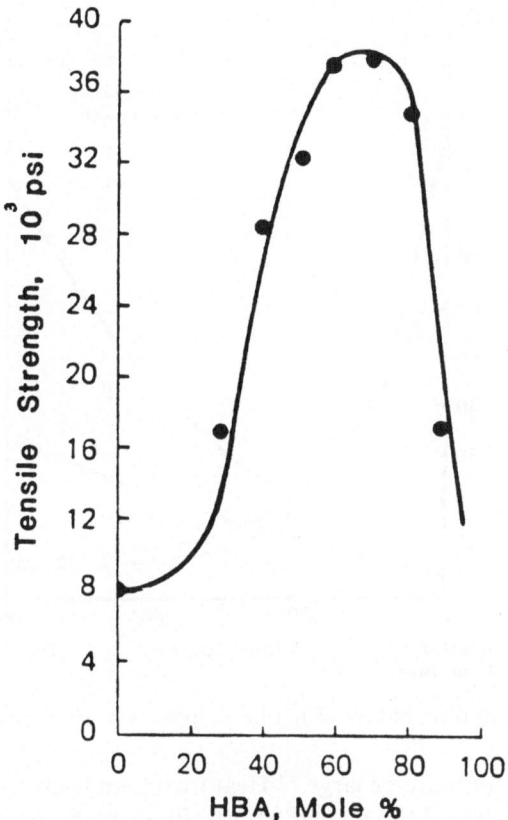

FIG. 14.  Effect of composition on the tensile strength of injection-molded bars of copoly(HBA/PET).[24]

is thought to be akin to a glass transition, and can be minimized by annealing.[23]

### 3.2.2. Molded Plastics

Figure 14 illustrates how the tensile strength of injection-molded bars of the HBA/PET copolymer change as a function of composition.[24] Due to the anisotropic structure of these LC polymers, one should expect different properties in the transverse and flow directions. Figure 15 shows the mechanical property distribution in a LCPE film as a function of test angle with respect to the extrusion direction. This behavior is similar to that of continuous reinforced composites, where the ratio of the 0° to the

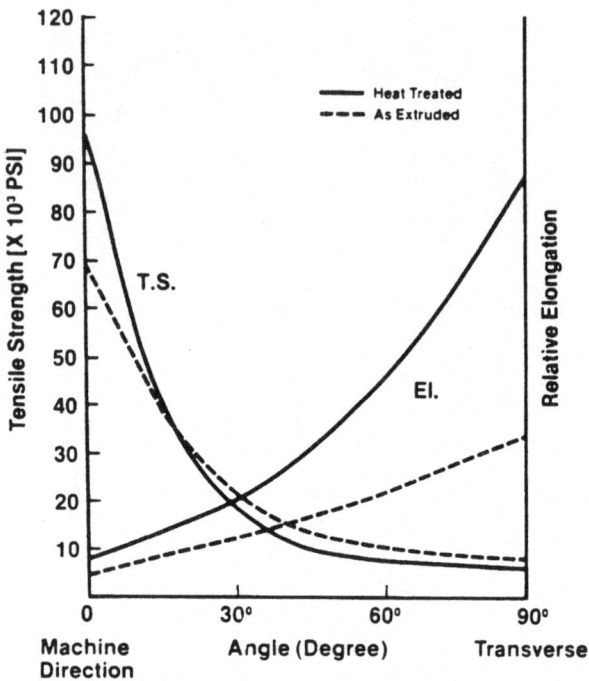

FIG. 15.   Angular distribution of tensile properties in an extruded LCPE film.[23]

90° tensile strength can be large.[23] Heat treatment leads to a higher tensile strength at 0°, with little change at 90°. Figures 16 and 17 illustrate the effect of thickness on the along- and across-the-flow properties.[24,48] Similar anisotropic properties are seen in fiber-reinforced composites and hence molded LCPEs are sometimes referred to as 'self reinforcing composites'. In addition, they exhibit good impact strengths, high heat deflection temperatures, and low thermal expansion coefficients, which are desirable properties for industrial moldings. Heat aging in air has virtually no effect on the properties.[23] At present, the possibilities for blending LCPEs with commercial thermoplastics like PET and PVC are being examined.[49,50]

### 3.3. Other Properties

The resistance of LCPEs to chemical degradation is illustrated in Table 4. The retention of properties in both acidic and basic environments is very good.[23] Unlike polyamides, water uptake is minimal, and these polymers

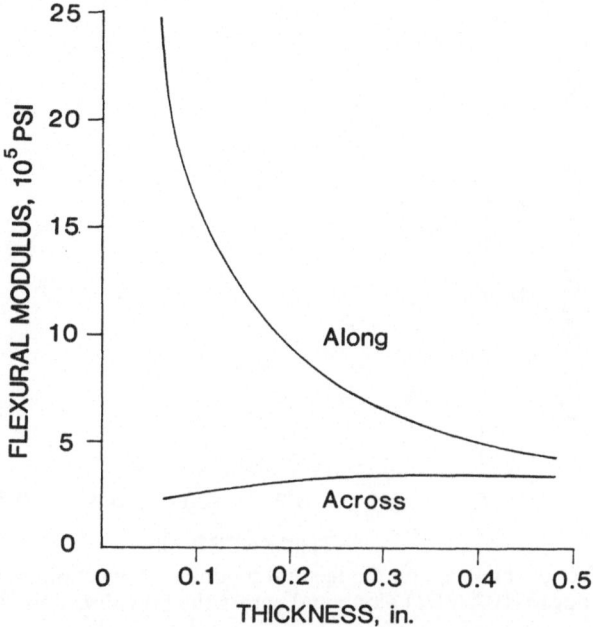

FIG. 16. Effect of thickness on the flexural strength of injection-molded bars of copoly(HBA/PET) along and across the flow direction.[24]

TABLE 4

CHEMICAL RESISTANCE OF AN LCPE FIBER[23] (MODULUS AND STRENGTH RETENTION OF FIBERS AGED IN SOLVENTS FOR 1 MONTH AT 50°C)

| Solvent | HBA/HNA | |
|---|---|---|
| | Mod. ret. (%) | Str. ret. (%) |
| Original | 505 | 21·4 |
| Water | 103 | 106 |
| Gasoline | 96 | 95 |
| Motor oil | 96 | 89 |
| 50% Antifreeze | 103 | 98 |
| 5% Sodium hypochlorite | 84 | 75 |
| 10% Sodium hydroxide | 88 | 40 |
| 20% Sulfuric acid | 94 | 88 |
| 20% Hydrochloric acid | 108 | 117 |

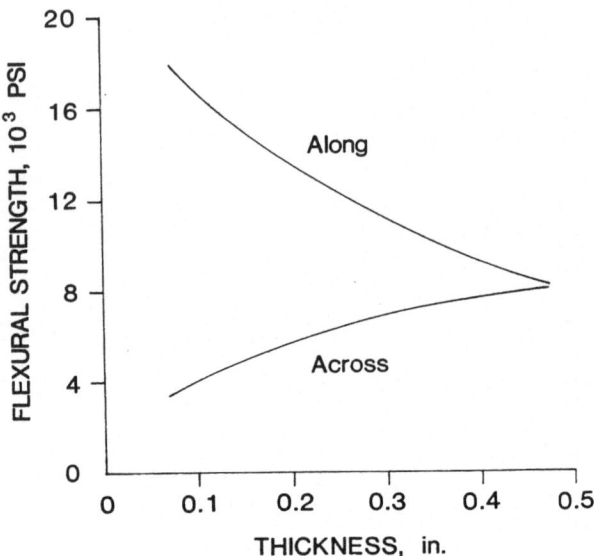

FIG. 17.    Effect of thickness on the flexural modulus of injection-molded bars of copoly(HBA/PET) along and across the flow direction.[24]

are insoluble in common laboratory solvents. Characterization in solution requires the use of relatively exotic solvents, such as pentafluorophenol, trifluoromethanesulfonic acid and chlorophenol, sometimes at high temperatures.

Many of the optical properties of LCPEs are a direct consequence of the anisotropy. Stir opalescence of the melt has been used to study the onset or disappearance of liquid crystalline phases.[30] Various optical textures, e.g. schlieren patterns, can be observed under the polarizing microscope, and these are the basis for classifying the different LC phases.[51] Nematic melts can be aligned by strong electric and magnetic fields. Electric fields can also induce instabilities and turbulence patterns in the melt,[30] although the technological importance of these effects has yet to be established. LCPEs have good electrical insulating properties (high dielectrical strength and arc resistance). They also have good burn resistance with minimal smoke generation, and are resistant to high levels of UV, ionizing and cobalt-60 radiation.[52]

In summary, LCPEs possess a wide array of desirable properties which can be raised to the levels of high performance composites and metals by proper processing. The control of the inherent anisotropy is the key to achieving the best results from these systems.

## 4. MORPHOLOGY AND STRUCTURE OF LC COPOLYESTERS

Most LCPEs exhibit a characteristic fibrous, almost wood-like texture, with a high surface gloss.[23] Fracture and etched surfaces of fibers and molded bars reveal a layered internal morphology. Transverse sections of injection molded plaques of copoly(HBA/HNA) show a skin–core delineation. Under the electron microscope, the etched transverse sections reveal poorly bonded layered sheets, whose thickness decreases towards the skin.[55] STEM and microdiffraction studies on microtomed sections of the copoly(HBA/PET) fiber showed a fibrillar or lamellar texture with poor section adhesion.[53] Heat treatment resulted in the definition of a biphasic structure, and it has been suggested that this is made up of an oriented phase of HBA-rich sequences in an amorphous PET-rich phase. X-ray studies of copoly(HBA/PET) by Blackwell et al.[54] also point to the existence of ordered regions with a structure similar to that of the high temperature form of poly(HBA). DSC and NMR data[56–59] have also been interpreted in terms of a biphasic structure, in which the ordered phase is rich in HBA.

Molecular orientation in the sheared melts of LCPEs have been studied by Viney et al.[60] They have observed banded structures under the optical microscope, which are believed to be the consequence of a periodic variation in the orientation of the principal optical vibration direction about the shear axis. This in turn is related to the corresponding variation in the direction of the long axis of the molecule. They also predict the presence of rotational correlations about the long axis of adjacent molecules to explain the optical behavior under crossed polars.[61] Orientation in the solid state has been studied using X-ray diffraction techniques. Mitchell and Windle[62,63] have analyzed the scattering patterns obtained from aligned samples of copoly(HBA/PET), and the derived values of the orientation parameters suggest the presence of local rotational correlation.

The limited solubility of most LCPEs has naturally restricted analyses of the solution properties. Calundann et al.[23] have studied the intrinsic viscosity–molecular weight relationships of a group of wholly aromatic copolyesters using GPC–LALS techniques and 50/50 hexafluoroiso-propanol–pentafluorophenol at 25°C as solvent. They have reported a typical value of $a = 0.98$ for the Mark–Houwink coefficient, for molecular weights between 5000 and 50 000, suggesting a semi-rigid conformation in solution. Copoly(HBA/PET) containing up to 40% HBA is soluble in chloroform; preparations containing up to 70% HBA are soluble in phenol–tetrachloroethane mixtures.[24] NMR analyses for these copolymers indicate a random monomer sequence; the calculated preference factor for HBA–HBA linkages had an upper limit of 1·3 at higher HBA contents.[56] At HBA contents above 70%, these copolymers are insoluble, and it has not been possible to investigate the suggestion that the reported biphasic structure is enhanced by non-random sequence distributions.

## 5. X-RAY ANALYSES OF THE STRUCTURE OF WHOLLY AROMATIC COPOLYESTERS

X-ray diffraction patterns of melt-spun fibers of wholly aromatic LC copolyesters prepared from HBA and HNA, and from HBA, 2,6-dihydroxynaphthalene (DHN) and TPA are shown in Plates 1–6 (see pp. 183–185) for three monomer ratios of each copolymer. The patterns reveal a high degree of axial orientation of the molecules, and the presence of sharp equatorial and off-equatorial Bragg maxima points to the existence of some three-dimensional order. The most striking feature is the presence of meridional maxima which are aperiodic, i.e. they are not orders of a simple repeat, and also shift progressively with monomer ratio.[64] Figure 18 shows the meridional diffractometer scans for five different monomer ratios of the HBA/HNA copolymer. The position and intensity of the peak at $2\theta \approx 43°$ ($d \approx 2·1$ Å) are approximately independent of composition, but the maximum in the region of $2\theta \approx 12°$ ($d \approx 7$ Å) moves steadily to lower angles and the maximum at $2\theta \approx 30°$ ($d \approx 2·9$ Å) moves steadily to higher angles as the HBA content decreases. At approximately 50% HBA content, a fourth maximum appears at $2\theta \approx 22°$ ($d \approx 4·1$ Å) which maintains its position but gets progressively more intense as the HBA content is decreased further. Analogous shifts in the positions of the meridional maxima with composition also occur in the HBA/DHN/TPA terpolymer.[65] Table 5 lists the positions of the meridional maxima for the HBA/HNA and HBA/DHN/TPA systems at three different compositions. It should be noted that no analytical data are currently available on the monomer sequence distribution in wholly aromatic copolyesters. Application of NMR methods, for example, is difficult because of the low solubility and the chemical similarity of the monomers. However, the above data argue against the existence of extensive block copolymer structure, and our analysis commenced at the other extreme, with a model consisting of a nematic assembly of parallel chains of completely random monomer sequence with poor lateral packing.[66]

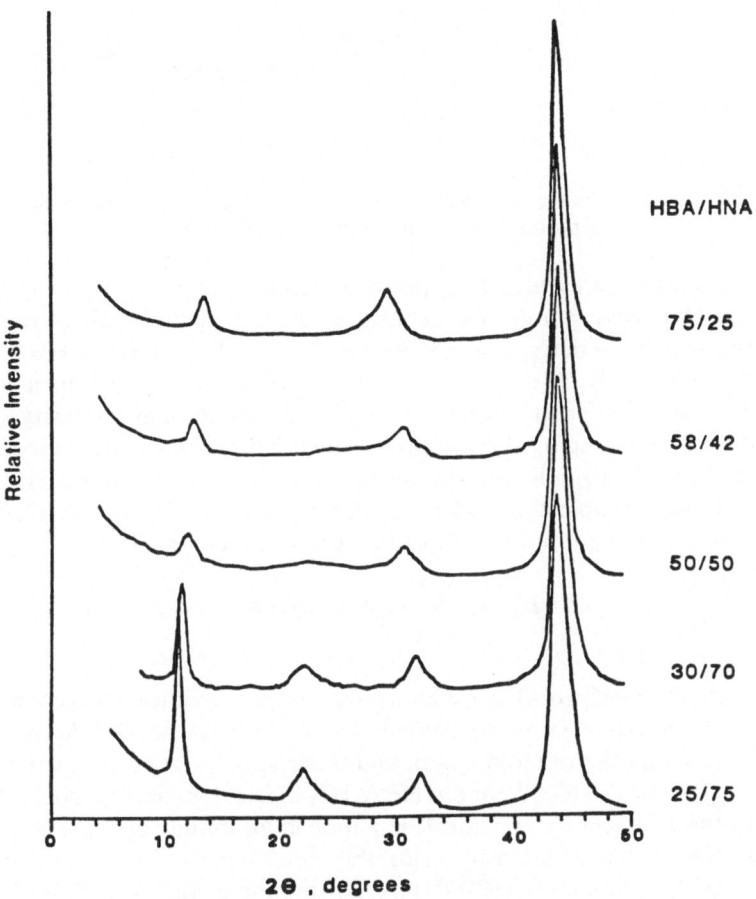

HBA/HNA

75/25

58/42

50/50

30/70

25/75

**Relative Intensity**

0      10      20      30      40      50

**2θ . degrees**

FIG. 18.    Meridional $\theta/2\theta$ diffractometer scans for five different monomer ratios of copoly(HBA/HNA).[64]

Figure 19 shows a projection of a model of a typical random sequence of copoly(HBA/HNA), constructed using standard bond lengths and angles and assuming planarity of the aromatic and ester groups. Torsional rotations about the aromatic–carboxyl linkage represent the only conformational freedom, and these torsion angles are set to $\pm 30°$ or $\pm 150°$, consistent with the results of conformational analysis on similar polyesters.[67,68] Since the aromatic–carboxyl bonds are approximately parallel to the fiber axis, the z-projection of the chain will be nearly independent of the conformation. As the first approximation, each

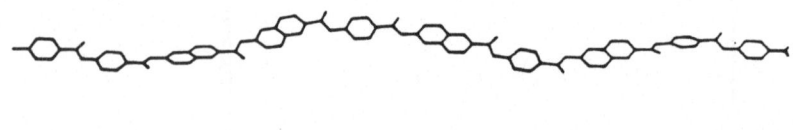

FIG. 19. A model chain of copoly(HBA/HNA) with a typical random sequence
of monomers and its point representation.[64]

monomer was represented by a point placed for convenience at the ester
oxygen. Adjacent points are separated by the corresponding residue
lengths, which are derived as 6·35 and 8·37 Å for the models of HBA and
HNA, respectively. Monte Carlo methods were used to set up a large
number of copolymer sequences, and the meridional intensity was
calculated by averaging their squared Fourier transforms. This approach
was successful in predicting the meridional maxima,[69,70] but it is more
efficient and complete to calculate the intensity, $I(Z)$, by the Fourier
transform of the autocorrelation function of the chain, $Q(z)$:

$$I(Z) = \sum_1 Q(z) \exp(2\pi i Z z_1) \tag{2}$$

where $Z$ is the reciprocal space coordinate in the direction corresponding
to the chain axis. $Q(z)$ is the probability of the 1st, 2nd, 3rd etc nearest
neighbor along the random chain, and is zero except at specific values of
$z = z_1$. $Q(z)$ for the 58/42 copoly(HBA/HNA) is shown in Fig. 20, plotted
out to the 14th (positive) nearest neighbor in an infinite chain. Figure 21
shows the calculated intensity using the above equation for chains of 15
point residues of copoly(HBA/HNA) for monomer ratios in 10% (mole)
increments over the entire composition range. As one goes from 100%
HBA to 100% HNA, the number of maxima increases from 3 to 4. (The
subsidiary maxima around the origin and $2\theta \approx 42°$ peaks are due to the
finite chain length.) The shifts in the peak position show trends similar to
those observed in the diffractometer traces in Fig. 18: the peaks at $2\theta \approx 12°$
and $2\theta \approx 30°$ move further apart and a new peak appears at $2\theta \approx 22°$ at a
mole ratio of 50%, and increases in intensity as the HBA content
decreases; meanwhile the peak at $2\theta \approx 44°$ remains constant across the
entire series.[71]

Good agreement between the observed and the calculated data is
obtained for the peak positions for both HBA/HNA and
HBA/DHN/TPA systems (see Table 5); it can be seen that the agreement is

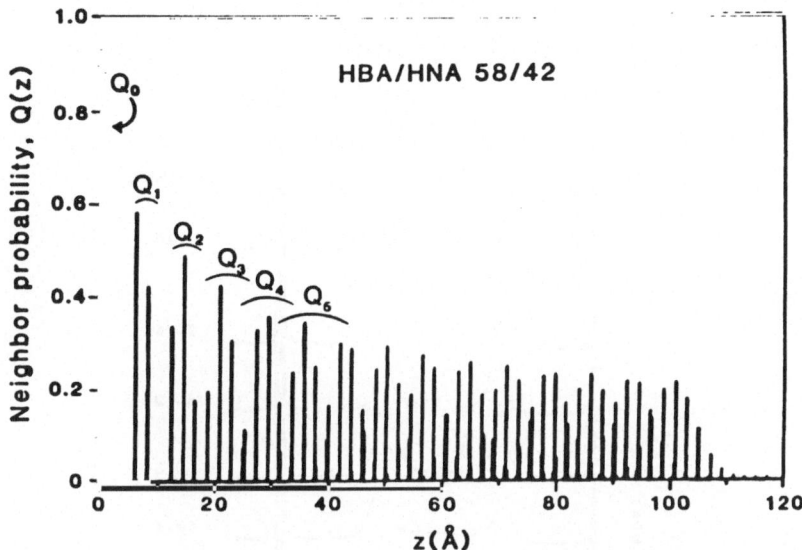

FIG. 20. Autocorrelation function for a point model of 58/42 copoly(HBA/HNA), plotted out to the 14th nearest neighbor; the first few neighbor contributions are shown as $Q_1$, $Q_2$ etc.

within $0.1$ Å in most cases. Note, however, that the relative intensities cannot be expected to match since intra-residue interferences are ignored when the residues are approximated as points.

The summation in eqn (2) has a closed solution which can be derived by treatment of the chain as a one-dimensional paracrystal, following the treatment by Hosemann.[72] For an infinite chain of point monomers, the intensity distribution has the simple form[73]

$$I(Z) = \text{Re} \left[ \frac{1 + H(Z)}{1 - H(Z)} \right] \qquad (3)$$

where $H(Z)$ is the Fourier transform of the 1st nearest neighbor distribution, $Q_1(z)$. A similar analysis has been used by Mitchell and Windle[74] to study the variation of the meridional intensity with composition. Previous treatments of the interference effects of one-dimensional lattices with two possible monomer repeats have been described by Hendricks and Teller in 1942 for layer structures such as clays[75] and by Bonart and Spei in 1972 for deformed $\alpha$-keratin.[76] An insight to the origin of the peaks and their composition dependence can

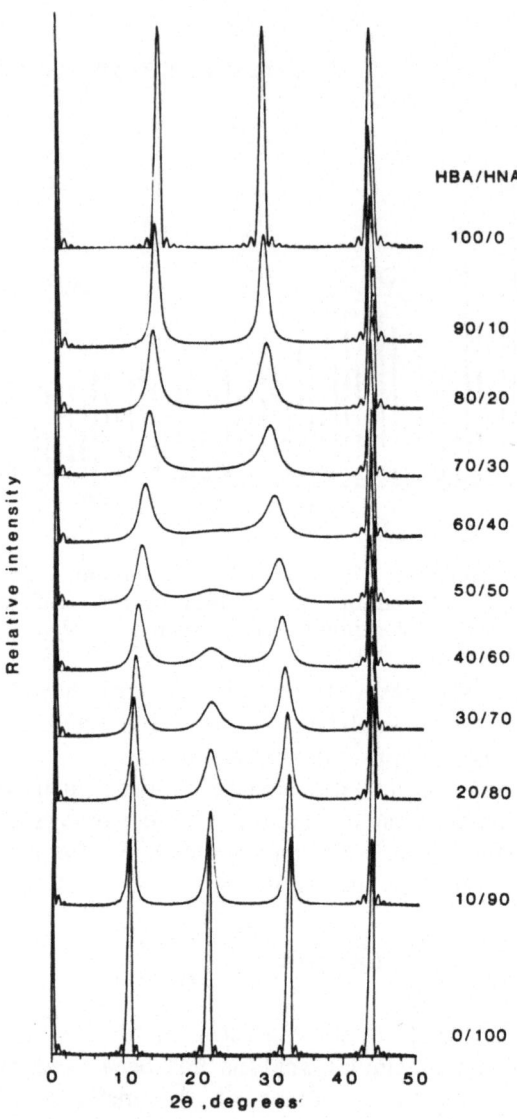

FIG. 21.  Point model calculations of the meridional intensity distribution over the entire composition range for copoly(HBA/HNA).[71]

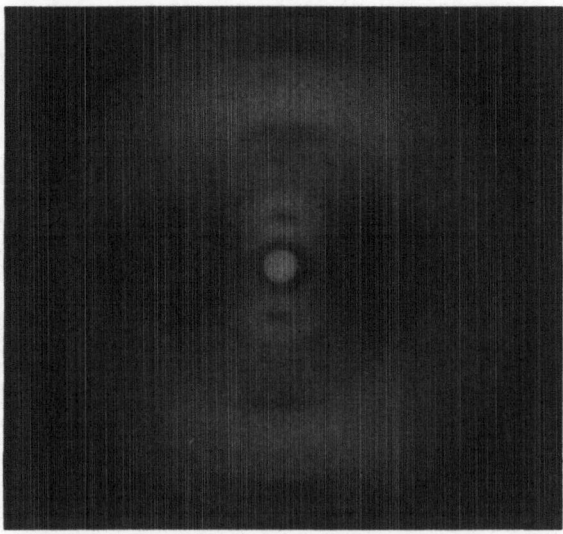

PLATE 1.  X-ray fiber diffraction pattern for copoly(HBA/HNA) monomer ratio of 75/25 (reproduced from Reference 70).

PLATE 2.  X-ray fiber diffraction pattern for copoly(HBA/HNA) monomer ratio of 58/42 (reproduced from Reference 70).

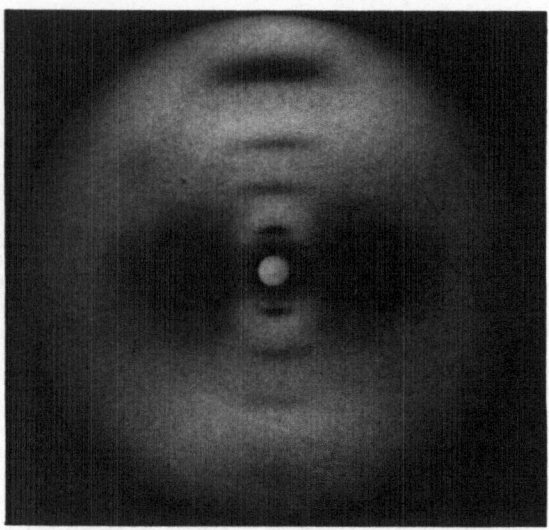

PLATE 3.  X-ray fiber diffraction pattern for copoly(HBA/HNA) monomer ratio of 30/70 (reproduced from Reference 70).

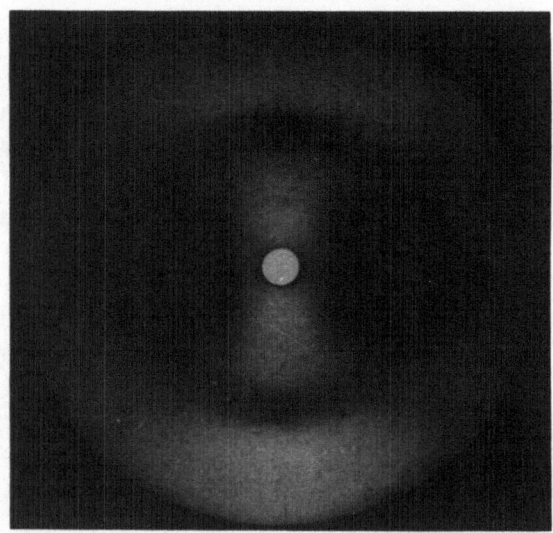

PLATE 4.  X-ray diffraction pattern from melt-drawn fibers of copoly(HBA/ DHN/TPA) for monomer ratio of 60/20/20 (reproduced from Reference 69).

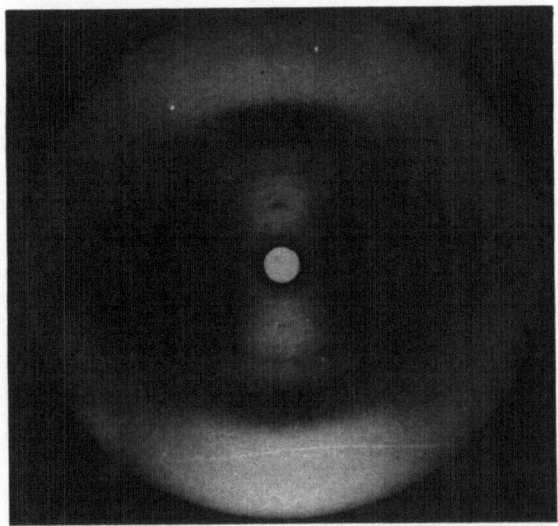

PLATE 5. X-ray diffraction pattern from melt-drawn fibers of copoly(HBA/ DHN/TPA) for monomer ratio of 50/25/25 (reproduced from Reference 69).

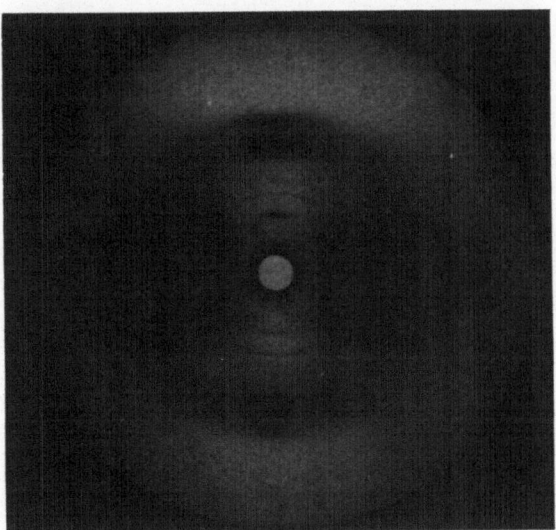

PLATE 6. X-ray diffraction pattern from melt-drawn fibers of copoly(HBA/ DHN/TPA) for monomer ratio of 40/30/30 (reproduced from Reference 69).

## TABLE 5

COMPARISON OF THE $d$-SPACINGS OF THE PEAKS ALONG THE MERIDIAN IN THE OBSERVED DATA WITH THOSE CALCULATED BY THE POINT AND ATOMIC MODELS FOR FINITE CHAINS

| Copolymer | Composition | $d$-spacings ($\mathring{A}$) | | |
|---|---|---|---|---|
| | | Observed | Point model | Atomic model |
| HBA/HNA | 75/25 | 6·78 | 6·75 | 7·04 |
| | | 3·03 | 3·09 | 3·09 |
| | | 2·03 | 2·11 | 2·11 |
| | 58/42 | 7·35 | 7·19 | 7·45 |
| | | | 4·01 | 4·40 |
| | | 2·98 | 2·98 | 2·99 |
| | | 2·05 | 2·10 | 2·11 |
| | 30/70 | 7·95 | 7·88 | 8·01 |
| | | 4·11 | 4·17 | 4·21 |
| | | 2·83 | 2·85 | 2·86 |
| | | 2·06 | 2·10 | 2·10 |
| HBA/DHN/TPA | 60/20/20 | 6·80 | 6·81 | 6·46 |
| | | 6·05 | | |
| | | 3·31 | 3·38 | 3·31 |
| | | 2·98 | 3·11 | 3·11 |
| | | 2·01 | 2·13 | 2·13 |
| | 50/25/25 | 6·98 | 6·95 | 6·54 |
| | | 5·90 | | |
| | | 3·38 | 3·47 | 3·42 |
| | | 2·97 | 3·08 | 3·09 |
| | | 2·02 | 2·13 | 2·13 |
| | 40/30/30 | 6·96 | 7·05 | 6·62 |
| | | 3·48 | 3·56 | 3·50 |
| | | 2·96 | 3·06 | 3·07 |
| | | 2·00 | 2·13 | 2·13 |

also be obtained by using phase-amplitude diagrams and deriving analytic equations for an 'average' phasor.[77]

To consider not only the positions of the aperiodic maxima but also their intensities, it is necessary to consider an atomic model for the chain, so as to allow for the intra- as well as the inter-residue interferences. This is achieved by separation of $Q(z)$ into its components:

$$Q(z) = Q(0) + \sum_A \sum_B Q_{AB}(z) \tag{4}$$

where $Q_{AB}(z)$ describes the probability of sequences beginning with residue A and ending with residue B. (For copoly(HBA/HNA), there will be four such $Q_{AB}(z)$'s.) The origin term is

$$Q(0) = \sum_A p_A \tag{5}$$

where $p_A$ is the mole fraction of monomer A. Hence we can write[78]

$$I(Z) = \sum_A p_A F_{AA} + \sum_A \sum_B F_{AB}(Z) \, \mathscr{F}[Q_{AB}(z)] \tag{6}$$

$$= \sum_A p_A F_{AA} + \sum_A \sum_B \sum_1 Q_{AB}(z) F_{AB}(Z) \exp(2\pi i Z z_1) \tag{7}$$

where $\mathscr{F}$ denotes the Fourier transform and $F_{AB}(Z)$ is the Fourier transform of the convolution of residue A with residue B:

$$F_{AB}(Z) = \sum_j \sum_k f_{j,A} f_{k,B} \exp[2\pi i Z(z_{k,B} - z_{j,A})] \tag{8}$$

Here, $f$ is the atomic scattering factor and the subscripts $j$,A and $k$,B denote the $j$th atom of residue A and the $k$th atom of residue B, respectively. Closed forms analogous to eqn (3) for an infinite atomic model can be derived by expressing $Q_{AB}(z)$ as the sum of two convolutions:

$$Q_{AB}(z) = p_A \sum_0^\infty Q_n(z) * p_B \delta(z - z_B) + p_B \sum_{-\infty}^0 Q_n(z) * p_A \delta(z + z_A) \tag{9}$$

for which the Fourier transform is

$$\mathscr{F}[Q_{AB}(z)] = p_A\left[1 + \frac{H(Z)}{1 - H(Z)}\right]p_B \exp(2\pi i Z z_B)$$

$$+ p_B\left[1 + \frac{H^*(Z)}{1 - H^*(Z)}\right]p_A \exp[2\pi i Z(-z_A)] \qquad (10)$$

Thus $I(Z)$ for an atomic model has the following form:[79]

$$I(Z) = \sum_A p_A F_{AA} + \sum_A \sum_B p_A 2\mathrm{Re}\left[\frac{F_{AB}(Z)p_B \exp(2\pi i Z z_B)}{1 - H(Z)}\right] \qquad (11)$$

Consideration of an atomic model leads to the prediction of maxima in approximately the same positions as in the point model (see Table 5) but now a reasonably good agreement of the relative intensities with the observed data is also achieved. Figure 22 compares the observed and calculated intensities for 30/70 copoly(HBA/HNA); the calculated $I(Z)$ data are corrected for the Lorentz/polarization effect. The intensity agreement can be seen to be good, and is typical of that obtained for the other monomer ratios. For the 30/70 copolymer the model is excellent for

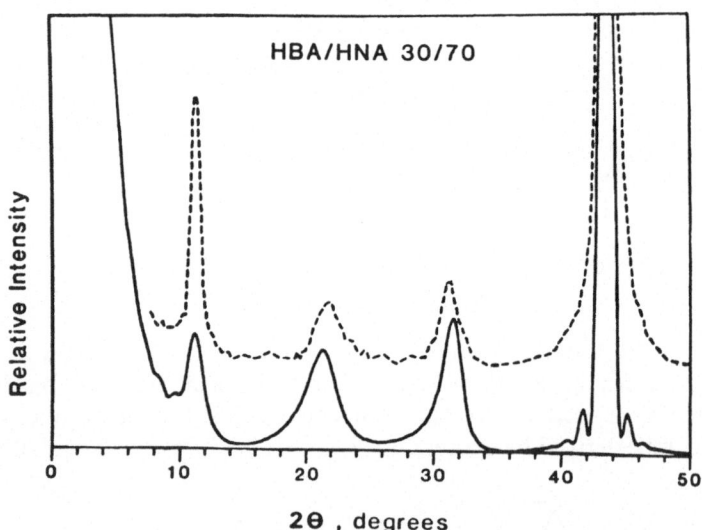

FIG. 22.   Comparison of the diffractometer data (dashed line) with the atomic model data (solid line) along the meridian for 30/70 copoly(HBA/HNA).[64]

the maxima at $d \approx 4 \cdot 1$, $2 \cdot 9$ and $2 \cdot 1$ Å. However, the peak at $d \approx 7$ Å has a higher relative intensity than that calculated. A similar discrepancy for the first maximum occurs for the other compositions, and our analyses suggest that our model is too 'perfect' in that we have assumed that all of the residues have their ester oxygen–ester oxygen vectors parallel to the $z$-axis. This can only be an approximation, as is apparent from Fig. 19; in

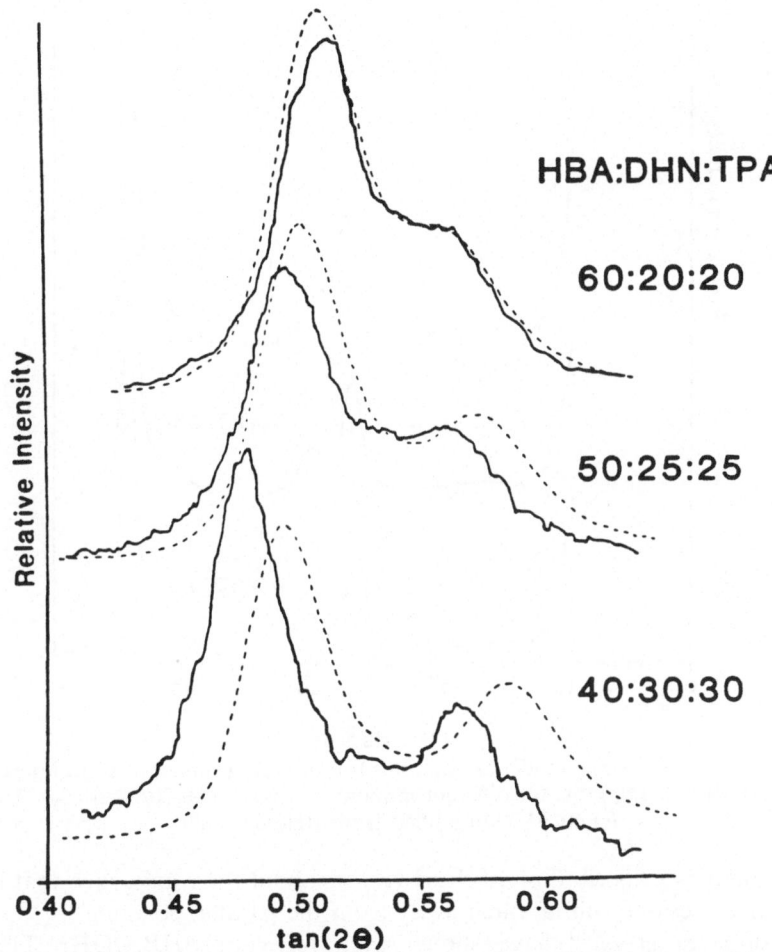

FIG. 23. Comparison of the atomic model data (dashed line) with the optical densitometer scan in the 3 Å region along the meridian for copoly-(HBA/DHN/TPA); the tilts for the HBA, DHN and TPA residues were set to 5°, 25° and 10° respectively.[80]

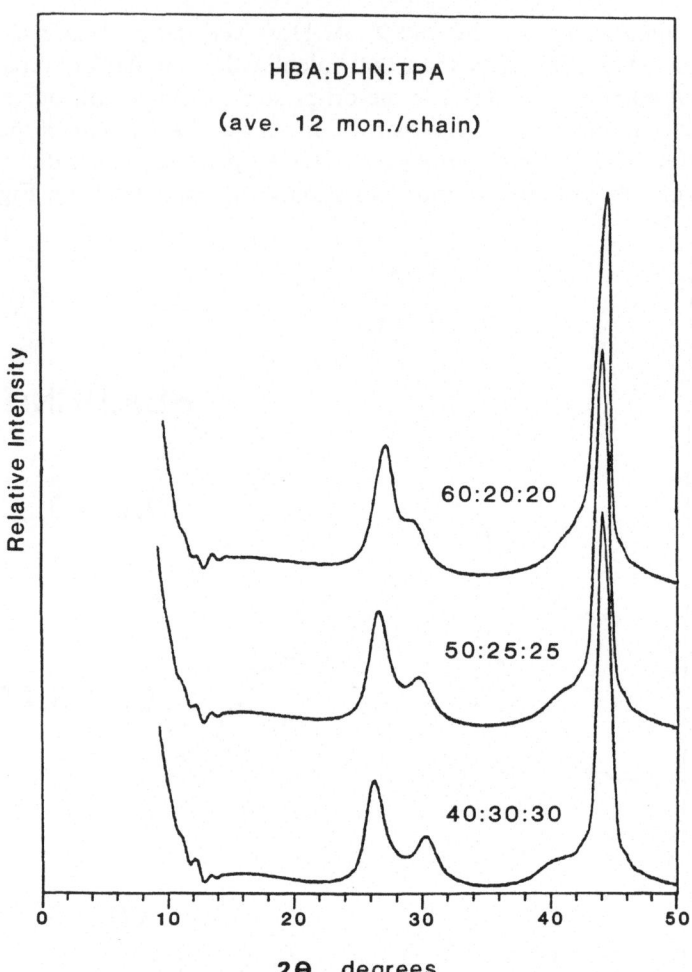

FIG. 24.   Tilted atomic model calculations for an average chain of 12 residues of copoly(HBA/DHN/TPA) for different monomer ratios: (a) 60/20/20; (b) 50/25/25; (c) 40/30/30 (modified from Reference 80).

fact there is a distribution of these vectors about the $z$-axis, such that the 'average' residue will be tilted away from the parallel position.

Gutierrez *et al.*[80] have shown that, for copoly(HBA/DHN/TPA), refinement of the 'average' residue orientations with respect to the fiber axis lead to a good match between the observed and calculated data. Figure 23 shows the calculated intensities for three monomer ratios which

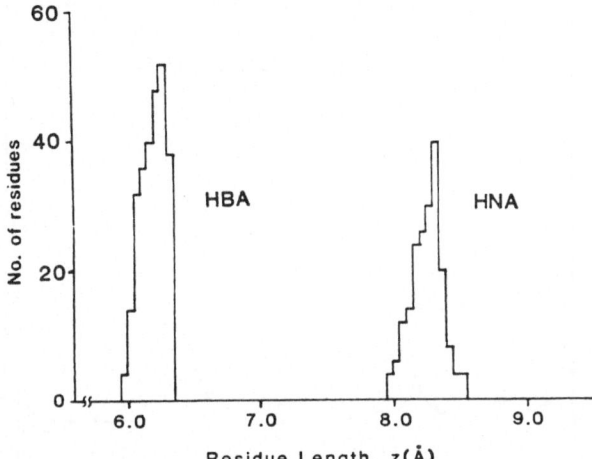

FIG. 25.   Residue length distribution for 58/42 copoly(HBA/HNA) obtained from 40 chains of 13 monomers each; the corresponding probability distribution is obtained by dividing by the total number of monomers.[81]

are in qualitatively good agreement in terms of the relative intensities in the fiber diagrams. The most striking of the data is the doublet at $d \approx 3$ Å, which increases in separation with decreasing HBA content. Figure 24 shows the agreement obtained between the calculated intensities and densitometer scans of the doublet (obtained from the X-ray patterns of fibers tilted so that $Z = 1/3\cdot1$ Å intersects the sphere of reflection).

The above refinement modelled the non-linearity of the chains by tilting each residue type at a constant angle to the chain axis. A more realistic approach is to incorporate a distribution of tilt angles. One can obtain a typical distribution of projected residue lengths by generating model chains such as those of copoly(HBA/HNA) (Fig. 19) from which we can construct a histogram of the residue lengths.[81] Figure 25 shows a residue length distribution for 58/42 copoly(HBA/HNA) from a survey of 40 chains of 13 residues each. The monomer sequences were obtained by the use of a random number generator, taking into account the monomer ratio with the torsion angles limited to $\pm30°$ and $\pm150°$ (selected at random). The distribution of residue lengths can be incorporated in the calculations for $I(Z)$ as follows:

$$I(Z) = \sum_{A} p_A F_{AA} + \sum_{A} \sum_{B} p_A 2\mathrm{Re}\left[\frac{F_{AB}(Z)\mathscr{F}[Q_B(z)]}{1 - H(Z)}\right] \quad (12)$$

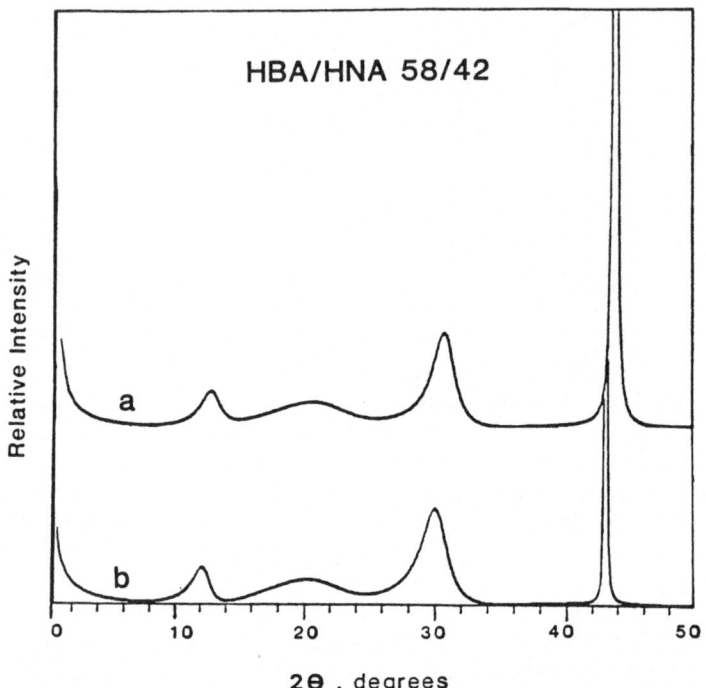

FIG. 26.    Meridional intensity distribution for 58/42 copoly(HBA/HNA): (a) for a
distribution of residue lengths; (b) for fixed residue lengths.

where $\mathscr{F}(Q_B(z))$ represents the Fourier transform over the individual
residue length distributions. Figure 26 compares $I(Z)$ calculated for 58/42
copoly(HBA/HNA) for a straight infinite chain (via eqn (11)) and a chain
incorporating the distribution of residue lengths via eqn (12).[81] It can be
seen that the results are very similar except that the peak at $d \approx 2 \cdot 1$ Å
broadens when the length distribution is incorporated. There is also a
general shift of the peak positions towards lower $d$ (higher $2\theta$) which is
explained by the fact that the average residue lengths must decrease when
we use a distribution of lengths rather than the maximum length in the
straight chain model.

The sharpness of the peak at $d \approx 2 \cdot 1$ Å in the straight chain model can
be understood in that this is a Bragg peak that arises due to the fact that
the lengths of the HBA and HNA residues (6·35 and 8·37 Å) are
fortuitously in the ratio 3:4. Hence the observed breadth of the 2·1 Å peak
can be interpreted in terms of a crystallite size, and we have shown that the

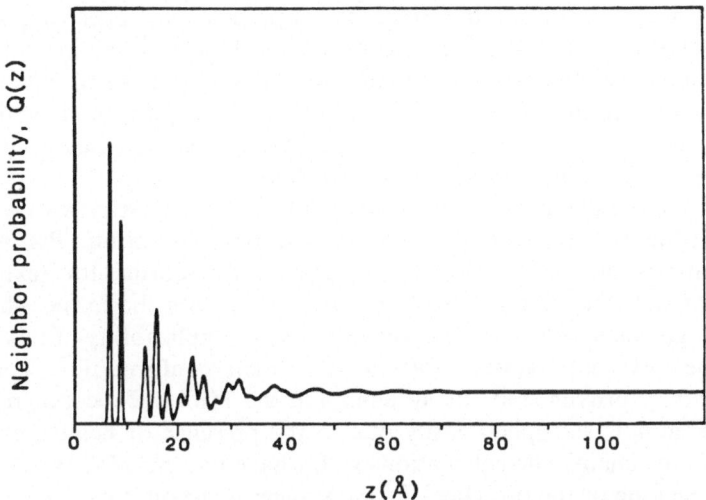

FIG. 27.   The autocorrelation function for 58/42 copoly(HBA/HNA) for a
distribution of residue lengths.

data for the 58/42 copolymer are reproduced by chains of $\sim 11$ monomers.[82] Since the molecular weights in the as-spun fibers are believed to be $\sim 25\,000$,[47] this chain length corresponds to a persistence or correlation length; it is the length beyond which the approximation of a straight chain breaks down.

Incorporation of the distribution of residue lengths based on a stereochemically acceptable model leads to a peak width at $d \approx 2{\cdot}1$ Å that is comparable to that observed. This match can be understood by incorporation of the length distribution function in Fig. 25 into the correlation function $Q(z)$ for the straight chain in Fig. 20. This combination leads to the correlation function in Fig. 27, which describes the distribution of nearest neighbors in chains such as those in Fig. 19. It can be seen that the peaks for successive neighbors get progressively broader until $Q(z)$ becomes smooth at a neighbor separation of $\sim 80$ Å, which corresponds to a length of $\sim 11$ residues.[82]

These analyses show that the X-ray scattering data are consistent with a highly extended chain configuration and a completely random monomer sequence. The question naturally arises as to how sensitive are the X-ray data to non-random sequence distribution. This can be modelled by modification of the $Q_{AB}(z)$ terms in eqn (4) to take account of the change in nearest neighbor probabilities as the sequence distribution becomes progressively non-random. Such an approach has been applied to copoly(HBA/DHN/TPA)[83] and copoly(HBA/HNA),[84] and in both cases the results show that all but minimal blockiness can be ruled out. Thus it appears that, in the special case of a stiff extended copolymer, in which the monomers have different axial lengths, X-ray methods can be used to investigate the monomer sequence distribution.

The above calculations yield a model for the copolyesters with which it is possible to consider the three-dimensional structure. Preliminary calculations on the cylindrically averaged transforms for (extended) random and rigid chains have been carried out, with the former showing better agreement with the observed data.[85] The applicability of this model to other LC copolyesters with a stiff chain conformation has been successfully proven, but for systems like the HBA/PET copolymer the model cannot be applied easily due to the presence of flexible ethylene units in the chain. The calculations so far have not included the effects of chain packing or the presence of axial stagger to account for the existence of three-dimensional order, and further work is in progress in this area.


## ACKNOWLEDGEMENT

This research was supported by NSF Grant No. DMR81-07130 from the Polymer Program.


## REFERENCES

1. REINITZER, F. (1888). *Monatsh. Chem.*, **9**, 421.
2. GRAY, G. W. (1962). *Molecular structure and the properties of liquid crystals*, Academic Press, London.
3. CHANDRASEKHAR, S. (1977). *Liquid crystals*, Cambridge University Press, London and New York.
4. SAEVA, F. D. (Ed.) (1979). *Liquid crystals: the fourth state of matter*, Marcel Dekker, New York.
5. BROWN, G. H. and WALKEN, J. J. (1979). *Liquid crystals and biological structure*, Academic Press, New York.
6. SAMULSKI, E. T. (1982). *Physics Today*, **35**(5), 40.
7. KWOLEK, S. L. (1972). US Patent 3671542 (E.I. DuPont de Nemours and Co.).
8. ECONOMY, J. and NOWAK, B. E. (1973). US Patent 3759870 (Carborundum Co.).
9. KUHFUSS, H. F. and JACKSON JR, W. J. (1974). US Patent 3804805 (Eastman Kodak Co.).

10. BLUMSTEIN, A. (Ed.) (1978). *Liquid crystalline order in polymers*, Academic Press, New York.
11. CIFERRI, A., KRIGBAUM, W. R. and MEYER, R. B. (Eds) (1982). *Polymer liquid crystals*, Academic Press, New York.
12. BLUMSTEIN, A. (Ed.) (1978). *Mesomorphic order in polymers and polymerization in liquid crystalline media*, ACS Symp. Ser. 74, Washington.
13. BLUMSTEIN, A. (Ed.) (1985). *Polymeric liquid crystals*, Plenum Press, New York.
14. LIPATOV, Y. S., TSUKRUCK, V. V. and SHILOV, V. V. (1984). *Rev. Macromol. Chem. Phys.*, C24(2), 173.
15. WENDORFF, J. H., PERPLIES, E. and RINGSDORF, H. (1975). *Progr. Colloid Sci.*, 57, 272.
16. DEMUS, D., DEMUS, H. and ZASCHKE, H. (1976). *Flussige Kristalle in Tabellen*, VEB Verlag, Leipzig.
17. BLUMSTEIN, A. and HSU, E. C. in Reference 10, p. 105.
18. WENDORFF, J. H., FINKELMAN, H. and RINGSDORF, H. (1978). *J. Polym. Sci. Polym. Symp.*, 63, 245.
19. FLORY, P. J. (1956). *Proc. Roy. Soc.*, A234, 73.
20. DEMARZIO, E. A. (1961). *J. Chem. Phys.*, 35, 658.
21. DOBB, M. G. and McINTYRE, J. E. (1978). *J. Polym. Sci. Polym. Symp.*, 63, 67.
22. IRWIN, R. S. (1985). Symp. on *Carbon and other high performance fibers*, ACS meeting, April.
23. CALUNDANN, G. W. and JAFFE, M. (1982). Robert A. Welch Conferences on Chemical Research Proc. Synth. Polymers, p. 247.
24. JACKSON JR, W. J. and KUHFUSS, H. F. (1976). *J. Polym. Sci. Polym. Chem.*, 14, 2043.
25. CIFERRI, A., in Reference 11, p. 63.
26. LUISE, R. R. (1980). US Patent 4183895 (E.I. DuPont de Nemours and Co.).
27. CALUNDANN, G. W. (1980). US Patent 4219461 (Celanese Co.).
28. FAVSTRITSKY, N. A. (1980). US Patent 4238598 (Fiber Industries Inc.).
29. JACKSON JR, W. J. and MORRIS, J. C. (1980). US Patent 4201856 (Eastman Kodak Co.).
30. JIN, J. I., ANTOUN, S. and LENZ, R. W. (1980). *Brit. Polym. J.*, 12, 132.
31. JACKSON JR, W. J. (1983). *Macromolecules*, 16(7), 1027.
32. GRIFFINS, A. C. and HAVENS, S. J. (1981). *J. Polym. Sci. Polym. Phys.*, 19, 951.
33. FLORY, P. J. and RONCA, G. (1979). *Mol. Cryst. Liq. Cryst.*, 54, 289.
34. VAN LUYEN, D. and STRZELECKI, L. (1980). *Eur. Polym. J.*, 16, 299.
35. OBER, C., LENZ, R. W., GALLI, G. and CHIELLINI, E. (1983). *Macromolecules*, 16, 1034.
36. ROVIELLO, A. and SIRIGU, A. (1982). *Makromol. Chem.*, 183, 895.
37. BLACKWELL, J., NAGARAJON, M. R. and HOITINK, T. B. (1979). *Polymer*, 20, 13.
38. BLUNDELL, D. J. (1982). *Polymer*, 23, 359.
39. GRIFFIN, B. P. and COX, M. K. (1980). *Brit. Polym. J.*, 12, 147.
40. DEMARTINO, R. N. (1983). *J. Appl. Polym. Sci.*, 28, 1805.
41. ZHOU, Q. F., LENZ, R. W. and JIN, J. I. in Reference 13, p. 257.
42. WISSBURN, K. E. (1980). *Brit. Polym. J.*, 12, 163.
43. BOLOTNIKOVA, L. S., BILBIN, A. Y., YEUSEYER, A. K., PANOV, Y. N., SKOROKHODOV, S. S. and FRENKEL, S. Y. (1983). *Polym. Sci. USSR*, 25(10), 2455.

44. BAIRD, D. G., in Reference 13, p. 119.
45. TAKEUCHI, Y., YAMAMOTO, F. and YAMAKAWA, S. (1984). *Brit. Polym. J.*, **16**(7), 579.
46. COGSWELL, F. N. (1980). *Brit. Polym. J.*, **12**, 12.
47. CALUNDANN, G. W. (1979). US Patent 4161470 (Celanese Co.).
48. HUYNH-BA, G. and CLUFF, E. F., in Reference 13, p. 217.
49. JOSEPH, E. G., WILKES, G. L. and BAIRD, B. G., in Reference 13, p. 197.
50. JACKSON, W. J. and KUHFUSS, H. F. (1980). *J. Appl. Polym. Sci.*, **25**, 1685.
51. NOEL, C., in Reference 13, p. 21.
52. Plastiscope (1984). *Mod. Plastics*, Dec. **14**, 14.
53. SAWYER, L. C. (1984). *J. Polym. Sci. Polym. Lett.*, **22**, 347.
54. BLACKWELL, J., LEISER, G. and GUTIERREZ, G. A. (1983). *Macromolecules*, **16**, 1418.
55. WENG, T., personal communication.
56. NICELY, V. A., personal communication.
57. ACIERNO, D., LA MANTIA, F. P., POLIZOTTI, G., CIFERRI, A. and VALENTI, B. (1982). *Macromolecules*, **15**(6), 1455.
58. ZACHARIADES, A. E., ECONOMY, J. and LOGAN, J. A. (1982). *J. Appl. Polym. Sci.*, **27**, 2009.
59. MEESIRI, W., MENCZEL, J., GAUR, U. and WUNDERLICH, B. (1982). *J. Polym. Sci. Polym. Phys.*, **20**, 719.
60. VINEY, C. and WINDLE, A. H. (1982). *J. Mater. Sci.*, **17**, 261.
61. VINEY, C., MITCHELL, G. R. and WINDLE, A. H. (1983). *Polym. Commun.*, **24**, 145.
62. MITCHELL, G. R. and WINDLE, A. H. (1982). *Polymer*, **23**, 1269.
63. MITCHELL, G. R. and WINDLE, A. H. (1983). *Polymer*, **24**, 1513.
64. CHIVERS, R. A., BLACKWELL, J., GUTIERREZ, G. A., STAMATOFF, J. B. and YOON, H., in Reference 13, p. 153.
65. BLACKWELL, J., GUTIERREZ, G. A. and CHIVERS, R. A., in Reference 13, p. 167.
66. BLACKWELL, J., BISWAS, A., GUTIERREZ, G. A. and CHIVERS, R. A. (1985). *Faraday Discuss. Chem. Soc.*, **79**, paper 6.
67. ADAMS, B. J. and MORSI, S. E. (1976). *Acta Cryst.*, **B32**, 1345.
68. HUMMEL, J. P. and FLORY, P. J. (1980). *Macromolecules*, **13**, 479.
69. BLACKWELL, J. and GUTIERREZ, G. A. (1982). *Polymer*, **23**, 671.
70. GUTIERREZ, G. A., CHIVERS, R. A., BLACKWELL, J., STAMATOFF, J. B. and YOON, H. (1983). *Polymer*, **24**, 937.
71. BLACKWELL, J., GUTIERREZ, G. A. and CHIVERS, R. A. (1984). *Macromolecules*, **17**, 1219.
72. HOSEMANN, R. (1950). *Z. Physik*, **128**, 465.
73. BLACKWELL, J., GUTIERREZ, G. A., CHIVERS, R. A. and RULAND, W. J., (1984). *J. Polym. Sci. Polym. Phys.*, **22**, 1343.
74. MITCHELL, G. R. and WINDLE, A. H. (1985). *Colloid Polym. Sci.*, **263**, 230.
75. HENDRICKS, S. and TELLER, E. (1942). *J. Chem. Phys.*, **10**, 147.
76. BONART, R. and SPEI, M. (1972). *Kolloid-Z., Z. Polym.*, **250**, 385.
77. DAVIES, G. R. and JAKEWAYS, R. (1985). *Polym. Commun.*, **26**(1), 9.
78. CHIVERS, R. A., BLACKWELL, J. and GUTIERREZ, G. A. (1984). *Polymer*, **25**, 435.
79. BLACKWELL, J., BISWAS, A. and BONART, R. (1985). *Macromolecules*, **18**, 2126.
80. GUTIERREZ, G. A., CHIVERS, R. A. and BLACKWELL, J. (1985). *Polymer*, **26**, 348.

81. BISWAS, A. (1985). M.S. Thesis, Case Western Reserve University.
82. BLACKWELL, J., CHIVERS, R. A., GUTIERREZ, G. A. and BISWAS, A. (1985–86). *J. Macromol. Sci. Phys.*, **B24**(1–4), 39.
83. GUTIERREZ, G. A. and BLACKWELL, J. (1984). *Macromolecules*, **17**, 2744.
84. BISWAS, A. and BLACKWELL, J. *Macromolecules.* (Submitted for publication.)
85. CHIVERS, R. A. and BLACKWELL, J. (1985). *Polymer*, **26**, 997.

*Chapter 6*

# SEGMENTAL ORIENTATION AND CHAIN RELAXATION OF POLYMERS BY SPECTROSCOPIC TECHNIQUES: A MOLECULAR APPROACH TO POLYMER VISCOELASTICITY

LUCIEN MONNERIE

*Laboratoire de Physicochimie Structurale et Macromoléculaire,
Ecole Supérieure de Physique et de Chimie Industrielles
de la Ville de Paris, France*

## 1. INTRODUCTION

A recent renewal of interest in the viscoelasticity of entangled bulk polymers has arisen from the theoretical approach of chain relaxation developed by de Gennes[1] and Doi and Edwards.[2,3] These theories are based on a molecular description of chain motions which involves reptation of the chain along the tube formed by neighbouring chains.

Although such theories can be tested through their predictions of macroscopic viscoelastic properties of bulk materials, such as viscosity and elastic and loss moduli, it is very tempting to obtain direct information on chain dynamics on a molecular scale. For this purpose spectroscopic techniques such as small-angle neutron scattering, infrared dichroism (IRD) and fluorescence polarization (FP) are particularly suitable. Small-angle neutron scattering yields information either on the molecular dimensions of the whole chain or on the structure of part of the chain whereas infrared dichroism and fluorescence polarization applied to stretched samples lead to information about the chain segment orientation.

This chapter will deal exclusively with infrared dichroism and fluorescence polarization. The chain relaxation is analysed through its effect on the segmental orientation achieved during stretching above the glass transition temperature. After a short presentation of these techniques, their application to polystyrene samples stretched above their glass transition temperature will be considered. The influence of the stretching conditions and of the polymer molecular weight are examined. The experimental results will be compared with predictions of molecular viscoelasticity theories of polymer melts and an improvement of the involved molecular model will be presented.

## 2.  ORIENTATION DISTRIBUTION FUNCTION

Before considering the IRD and FP techniques which lead to measurement of the orientation distribution of characteristic vectors, some convenient quantities will be introduced to describe the orientation of uniaxially symmetric systems.

Each vector is characterized by the angle $\theta$ that its direction makes with the symmetry axis. The orientation distribution is represented by a function $f(\theta)$ which can be expanded in terms of Legendre polynomials in $\cos \theta$ as follows:

$$f(\theta) = \sum_l b_l P_l(\cos \theta) \tag{1}$$

with

$$b_l = (1/2\pi)(2l + 1)/2 \langle P_l(\cos \theta) \rangle$$

where $\langle P_l(\cos \theta) \rangle$ is the value of $P_l(\cos \theta)$ averaged over the distribution.

The orientation functions commonly used are defined as:

$$\langle P_2(\cos \theta) \rangle = (1/2)\langle 3 \cos^2 \theta - 1 \rangle \tag{2}$$

$$\langle P_4(\cos \theta) \rangle = (1/8)\langle 35 \cos^4 \theta - 30 \cos^2 \theta + 3 \rangle \tag{3}$$

where

$$\langle \cos^n \theta \rangle = \int_0^\pi f(\theta) \cos^n \theta \sin \theta \, d\theta$$

It has recently been shown[4] that for a uniaxial distribution the determination of $\langle P_2(\cos \theta) \rangle$ and $\langle P_4(\cos \theta) \rangle$ is sufficient in most cases.

## 3. INFRARED DICHROISM

### 3.1. Principle of Orientation Measurements

Absorption in the infrared region of the spectrum deals with vibrational motions of the various atoms of a molecule. This leads to a set of normal vibration frequencies which can be assigned to specific vibration modes of small groups of atoms (stretching, bending, twisting, etc.). Each mode has a transition moment, $M$, with a definite orientation in the molecule (Fig. 1). The intensity of an infrared absorption band depends upon the angle the electric vector, $E$, in the incident radiation makes with the transition moment $M$. The dichroic ratio, $R$, defined as

$$R = A_{\parallel}/A_{\perp}$$

($A_{\parallel}$ and $A_{\perp}$ being the measured absorbance for electric vectors parallel and perpendicular, respectively, to the preferential orientation axis), is related to the orientation function $\langle P_2(\cos\theta)\rangle$ by

$$\langle P_2(\cos\theta)\rangle = (1/2)\langle 3\cos^2\theta - 1\rangle = \left(\frac{R-1}{R+2}\right)\left(\frac{R_0+2}{R_0-1}\right) \qquad (4)$$

where $\theta$ is the angle between the molecular axis and the preferential orientation axis (Fig. 1) and $R_0 = 2\cot^2\alpha$, $\alpha$ being the angle between the transition moment $M$ associated with the considered absorption band and the molecular axis (Fig. 1).

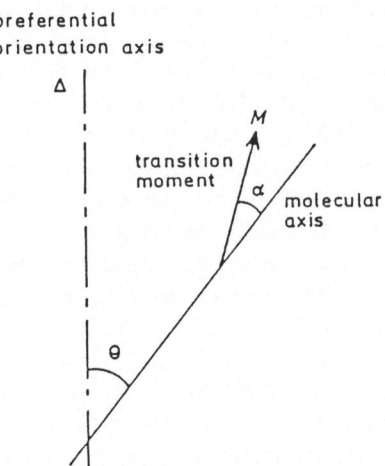

FIG. 1.  Position of molecular axis and transition moment with respect to the reference axis, $\triangle$.

## 3.2. Application to the Orientation of Polymers

Infrared dichroism has been applied for many years to orientation measurements involving polymers; these studies have been discussed.[5,6] Dichroism measurements are easily performed either on dispersive instruments or, more conveniently, on a Fourier transform apparatus. Measurements can be carried out either on stretched samples or during stretching. The main practical problem in the case of Fourier transform measurements arises from the requirement of band absorbance lower than *ca.* 0·7 absorbance units in order to permit use of the Beer–Lambert law. This means that one must obtain sufficiently thin films. Depending on the extinction coefficient of the considered band, the required thickness can range from 1 to 200 $\mu$m. From this point of view, polymers with strong absorption bands (e.g. polycarbonate) are difficult to study.

The main difficulty in using infrared dichroism to study orientation is the necessity to find infrared absorption bands of the polymer which are sufficiently well assigned to normal vibrations of specified atomic groups. Such an assignment can be achieved by making a normal-coordinate analysis and experimentally by looking at deuteration effects and dichroic behaviour. Furthermore, it is necessary that these well defined vibrational bands do not overlap with other bands resulting from another normal mode, a harmonic or a combination of other modes.

Another problem arises from the choice of a local chain axis which is convenient to describe the chain orientation. When dealing with vibration modes which are independent of the local conformation of the main chain (*trans* or *gauche*), the infrared dichroism will lead to the average orientation of chain segments, and the local chain axis must be chosen in such a way that the same value of the orientation function is obtained when various absorption bands of the same type, corresponding to different $\alpha$ values, are used to calculate $P_2$. An example of such a local chain axis is given in Fig. 2. Some vibration modes and absorption bands are characteristic of specific chain conformations (*ttt* for instance) and thus infrared dichroism will lead to the orientation value of these particular chain conformations. This can be done in the same way for absorption bands characteristic of the crystalline structure.

At this point, it is worth noting that the orientation function $\langle P_2(\cos\theta)\rangle$ obtained from IRD measurements corresponds to an average over all the groups of a chain involved in the considered vibration mode (independently of their position along the chain) and over all the chains of the same species in the sample; it will be called $\langle P_2(\cos\theta)\rangle_{av}$.

Finally, note that some experiments using deuterium-labelled

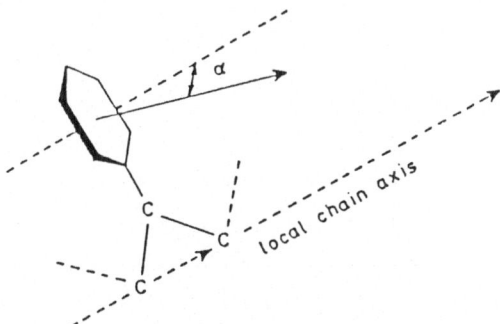

FIG. 2.   Local chain axis in polystyrene.

polystyrene chains blended with hydrogenated polystyrene have been performed recently by J. F. Tassin in our laboratory. Some specific bands do not overlap but allow measurement of the orientation of each species. Unfortunately, this method requires concentrations of deuterated chains in the blend greater than 10%, which makes it difficult to study the effect of the molecular weight of the labelled chains or of the polymer matrix. However, the study of tri-block copolymers in which, for example, the central block is hydrogenated whereas the other two are deuterated permits separate examination of the behaviour of the central part of the chain and of the ends. Such studies are at present in progress in our laboratory.

## 4.   FLUORESCENCE POLARIZATION

### 4.1. Principles

Fluorescent molecules have the property of re-emitting, in the form of visible light, part of the energy acquired by the absorption of luminous radiation. After illumination by a very short pulse at time $t_0$, the fluorescent light emitted at time $(t_0 + u)$ is proportional to $\exp(-u/\tau)$, where $\tau$ is the mean lifetime of the excited state (usually called the fluorescence lifetime). The most frequent $\tau$ values range from 1 to 100 ns. When absorbing light of a suitable wavelength, a molecule behaves as an electric dipole oscillator with a fixed orientation with respect to the geometry of the molecule. Such an equivalent oscillator is termed an absorption transition moment, $M_0$. In the same way, for the fluorescence emission we have an emission transition moment, $M$. When such a molecule receives an incident beam polarized along the $P$ direction (Fig. 3), the absorption probability is

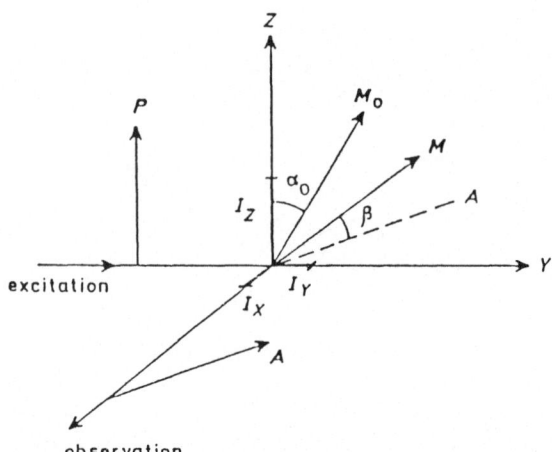

FIG. 3.    Polarized absorption and fluorescence emission: P, polarizer; A, analyser.

proportional to $\cos^2 \alpha_0$. In the same way, the fluorescence intensity measured through an analyser, $A$, is proportional to $\cos^2 \beta$. Thus for the $P$ and $A$ directions of polarizer and analyser the observed luminescence intensity is proportional to $\cos^2 \alpha_0 \cdot \cos^2 \beta$. Owing to the lack of phase correlation between excitation and emission lights, fluorescence emission can be described as resulting from three independent radiations respectively polarized along the $X$, $Y$ and $Z$ axes with intensities $I_X$, $I_Y$ and $I_Z$. The Curie symmetry principle, applied to excitation light polarized along $Z$, leads to $I_X = I_Y$. The fluorescence polarization is characterized by the emission anisotropy:

$$r = (I_{\|} - I_{\perp})/(I_{\|} + 2I_{\perp}) \tag{5}$$

where $I_{\|}$ and $I_{\perp}$ correspond to the fluorescence intensity obtained with an analyser direction parallel and perpendicular, respectively, to that of the polarizer.

### 4.2. Orientation of Uniaxially Symmetric Systems
For our present purpose we are mainly interested in the use of fluorescence polarization to look at the orientation distribution of fluorescent molecules. The main results which can be derived are presented below; more details can be found in a recent review[7] and in the original paper.[8]

In the following we will assume that the transition moments in both absorption and emission coincide with a molecular axis $M$ of the molecule, whose direction is specified by the spherical polar angle $\Omega = (\alpha, \beta)$ in the

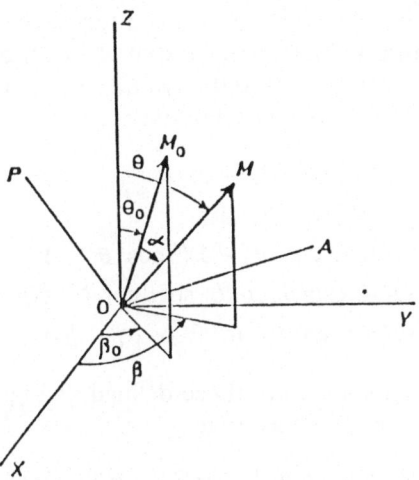

FIG. 4.  Illustration of the angles which define the orientation of molecular axis $M_0$ at time $t_0$ and $M$ at time $(t_0 + u)$ with respect to the fixed frame $OXYZ$.

reference frame (Fig. 4). Let us introduce the angular functions $N(\Omega_0, t_0)$, the orientation distribution of $M$ at time $t_0$ ($M_0$ in Fig. 4), and $P(\Omega, t/\Omega_0, t_0)$, the conditional probability density of finding at position $\Omega$ at time $t$ a vector $M$ which was at position $\Omega_0$ at time $t_0$.[9]

After illuminating the sample by a linearly polarized short pulse of light at $t_0$, the intensity emitted at time $(t_0 + u)$ for the $P$ and $A$ directions of polarizer and analyser is given by

$$i(P, A, t_0 + u) = K \int \int N(\Omega_0, t_0)P(\Omega, t_0 + u \,|\, \Omega_0, t_0)$$

$$\times \cos^2(P, M_0)\cos^2(A, M)\exp(-u/\tau)\,\mathrm{d}\Omega_0\,\mathrm{d}\Omega \quad (6)$$

where $K$ is an instrumental constant. In this expression $t_0$ corresponds to the macroscopic evolution of the sample, for example in a rheological experiment, whereas $u$ corresponds to a microscopic reorientational motion in the scale of the fluorescence lifetime $\tau$. In most cases the $t_0$ dependence of $N$ and $P$ can be ignored within the time $\tau$ (ca. $10^{-8}$ s), and the fluorescence intensity emitted under continuous excitation is given by

$$i(P, A, t_0, \tau) = \int_0^\infty i(P, A, t_0 + u)\,\mathrm{d}u \quad (7)$$

In the case of a uniaxial symmetric distribution of the molecular axes $M$, the intensities corresponding to the $P$ and $A$ directions lying along the fixed-frame axes ($Z$ corresponds to the symmetry axis) can be conveniently expressed through the following quantities:

$$G_{20}^0 = (1/2)\langle 3\cos^2\theta_0 - 1\rangle$$
$$G_{02}^0 = (1/2)\langle 3\cos^2\theta - 1\rangle$$
$$G_{22}^0 = (1/4)\langle(3\cos^2\theta_0 - 1)(3\cos^2\theta - 1)\rangle \tag{8}$$
$$G_{22}^1 = (9/16)\langle\sin\theta_0\cos\theta_0\sin\theta\cos\theta\cos(\beta - \beta_0)\rangle$$
$$G_{22}^2{}_{\text{i}} = (9/64)\langle\sin^2\theta_0\sin^2\theta\cos 2(\beta - \beta_0)\rangle$$

The various angles involved are those defined in Fig. 3 and the angular brackets denote an ensemble average.

$$i(Z, Z) = (K/9)(1 + 2G_{20}^0 + 2G_{02}^0 + 4G_{22}^0)$$
$$i(Z, X) = (K/9)(1 + 2G_{20}^0 - G_{02}^0 - G_{22}^0)$$
$$i(X, Z) = (K/9)(1 - G_{20}^0 + 2G_{02}^0 - 2G_{22}^0) \tag{9}$$
$$i(X, X) = (K/9)(1 - G_{20}^0 - G_{02}^0 + G_{22}^0 + 3G_{22}^2)$$
$$i(X, Y) = (K/9)(1 - G_{20}^0 - G_{02}^0 + G_{22}^0 - 8G_{22}^2)$$

It is worth noting that the quantities $G_{20}^0$ and $G_{02}^0$ represent the second moments of the distribution of vectors $\vec{M}_0$ and $\vec{M}$ respectively, and they describe the molecular orientation independently of molecular mobility. The functions $G_{22}^m$ depend on both orientation and mobility.

Further investigation is required to consider if, during the fluorescence lifetime, motions can occur or not. We will hereafter refer to these situations as 'mobile' or 'frozen' systems, respectively.

### 4.2.1. Uniaxial Frozen Systems

In such cases no molecular motion occurs during the fluorescence lifetime $\tau$, and the quantities $G_{20}^0 = G_{02}^0$ and $G_{22}^m$ can be rewritten with only two independent quantities, $\cos^2\theta$ and $\cos^4\theta$. All the information on the fluorescence intensities may be displayed in a $3 \times 3$ tensor $I$:

$$I = K \begin{vmatrix} (3/8)\langle\sin^4\theta\rangle & (1/8)\langle\sin^4\theta\rangle & (1/2)\langle\sin^2\theta\cos^2\theta\rangle \\ (1/8)\langle\sin^4\theta\rangle & (3/8)\langle\sin^4\theta\rangle & (1/2)\langle\sin^2\theta\cos^2\theta\rangle \\ (1/2)\langle\sin^2\theta\cos^2\theta\rangle & (1/2)\langle\sin^2\theta\cos^2\theta\rangle & \langle\cos^4\theta\rangle \end{vmatrix}$$
$$\tag{10}$$

This is identical to the result given previously.[10] The apparatus constant can be obtained from

$$K = \sum_i \sum_j I_{ij} = (8/3)I_{XX} + 4I_{XZ} + I_{ZZ} \qquad (11)$$

in such a way that the quantities $\langle \cos^2 \theta \rangle$ and $\langle \cos^4 \theta \rangle$ can be derived from fluorescence intensity measurements by the relations

$$\langle \cos^2 \theta \rangle = (I_{ZZ} + 2I_{XZ})/[(8/3)I_{XX} + 4I_{XZ} + I_{ZZ}] \qquad (12)$$

$$\langle \cos^4 \theta \rangle = I_{ZZ}/[(8/3)I_{XX} + 4I_{XZ} + I_{ZZ}] \qquad (13)$$

from which the orientation functions $\langle P_2(\cos \theta) \rangle$ and $\langle P_4(\cos \theta) \rangle$ can be calculated using expressions (2) and (3) respectively.

### 4.2.2. Uniaxial Mobile Systems

In this case both orientation and mobility contribute to the fluorescence polarization. Nevertheless such systems can be treated.[8] Indeed, if we assume that during the fluorescence lifetime ($10^{-10}$ to $10^{-7}$ s) the orientation distribution does not change, it results from the general expressions for $G$:

$$G_{20}^0 = G_{02}^0 = (1/2)\langle 3\cos^2 \theta_0 - 1 \rangle = \langle P_2(\cos \theta_0) \rangle \qquad (14)$$

On the other hand, for mobile systems, $G_{22}^0$ depends on both orientation and mobility, in such a way that $\langle P_4(\cos \theta_0) \rangle$ cannot be obtained. Furthermore, intensity measurements corresponding to polarizer and analyser directions along the $X_0$ and $Z$ axis are not sufficient for deriving $G_{20}^0$. The set of quantities $G_{20}^0$ and $G_{22}^m$ must be obtained from the measurement of five intensities, $i(P, A)$, corresponding to $P$ and $A$ orientations which are not contained in the same plane, excluding the use of only a straight-through optical arrangement.

Concerning the mobility, from the available $G_{20}^0$ and $G_{22}^m$ quantities, one can derive the following.

(a) The mean mobility amplitude, $M$:

$$M = \langle (3\cos^2 \alpha - 1)/2 \rangle = G_{22}^0 + (16/3)G_{22}^1 + (16/3)G_{22}^2 \qquad (15)$$

describing the motion performed during the considered time interval. Thus, for a Dirac-pulse excitation,

$$M(t) = \langle (3\cos^2 \alpha(t) - 1)/2 \rangle \qquad (16)$$

whereas, for a continuous excitation, one gets

$$M(\tau) = \int_0^\infty M(t)\exp(-t/\tau)\,\mathrm{d}t \qquad (17)$$

where $\tau$ is the fluorescence lifetime.

(b) Three angular correlation functions which contain information on the anisotropy of the motion.[8]

### 4.2.3. Effect of Electronic Delocalization

The effects of non-parallel absorption and emission moments on orientation measurements have been considered by several authors.[10,11] For uniaxial systems, an easy correction has been derived.[8] The delocalization results in the fact that the measured intensities $i(\vec{P},\vec{A})$ do not lead to the $G$ quantities defined above, but to $\gamma$ coefficients which are related to them by the relations

$$\gamma_{20}^0 = (5r_0/2)^{1/2}G_{20}^0 \qquad (18)$$
$$\gamma_{22}^m = (5r_0/2)G_{22}^m \qquad \text{with } m = 0, 1, 2$$

where $r_0$ is the fundamental emission anisotropy, which must be determined on an unoriented sample from an additional measurement of emission anisotropy.

### 4.2.4. Birefringence and Light-scattering Corrections

The birefringence effect occurs when the direction of either the polarizer or the analyser does not coincide with a principal direction of the refractive index tensor. It has been shown[8] that only the quantity $\gamma_{22}^1$ is modified and becomes

$$\gamma_{22}^1 = (5r_0/2)bG_{22}^1 \qquad (19)$$

where $b$ is a correction factor which can be experimentally determined.[12]

When dealing with crystalline polymers, light-scattering occurs and modifies the state of polarization of both the excitation and emitted light. The principle of a method of correction has been given,[11] and applied to straight-through measurements.[13]

## 4.3. Instrumentation for Measuring Orientation in Uniaxial Systems

Since FP is an optical technique, only samples that are sufficiently transparent can be studied, i.e. amorphous polymers of any thickness or thin films ($< 100\,\mu$m) of semi-crystalline polymers.

For systems in which the orientation does not change with time, successive fluorescence intensity measurements can be performed for the required polarizer and analyser directions, leading to a determination of $\langle P_2(\cos\theta)\rangle$ and $\langle P_4(\cos\theta)\rangle$.

Measurements are performed under continuous excitation. For frozen systems, a straight-through optical arrangement perpendicular to the sample is very convenient. Front illumination has been proposed[14] but the effect of refraction of the light inside the sample must be considered.[12] Measurements under an optical microscope have been performed either on stretched samples[11] or during stretching.[13]

When dealing with mobile systems more elaborate equipment is required and it has been developed in our laboratory.[12] This apparatus permits simultaneous measurements of the intensities required for determining orientation and mobility even during stretching. The optical system is shown in Fig. 5. In the case of polymers, mobility studies performed by fluorescence polarization on isotropic bulk polymers[15] have shown that significant motion during the fluorescence lifetime only occurs at temperatures more than 50°C above $T_g$. Thus polymers below this temperature range can be considered as frozen systems.

### 4.4. Chain Labelling

The use of fluorescence polarization to derive the orientation functions $\langle P_2(\cos\theta)\rangle$ and $\langle P_4(\cos\theta)\rangle$ of a set of fluorescent molecules requires that there is no energy transfer between the fluorescent molecules, implying a concentration of fluorescent species in the sample below 100 ppm. Thus the intrinsic fluorescence of a monomer unit cannot be used, e.g. phenyl groups in polystyrene.

In order to correlate unambiguously the orientation of the transition moment of the fluorescence molecule with that of the polymer chain, it is necessary to carry out covalent labelling. This can be achieved by performing an anionic polymerization and deactivating the living chains with 9,10-bis(bromomethyl)anthracene. In this way the resultant polymer contains a centrally located fluorescent group in which the transition moment lies along the local chain axis (Fig. 6). End-chain labelling can be obtained by using a monofunctional anthracene derivative. Such a method has been successfully applied in our laboratory to polystyrene and various polydienes.

A small amount ($\sim 0.1$–$1\%$ w/w) of labelled polymer is mixed in solution with normal polymer, in such a way that the final concentration of anthracene in the polymer sample would be around $10^{-6}$ w/w. After solvent drying, the samples are pressure moulded.

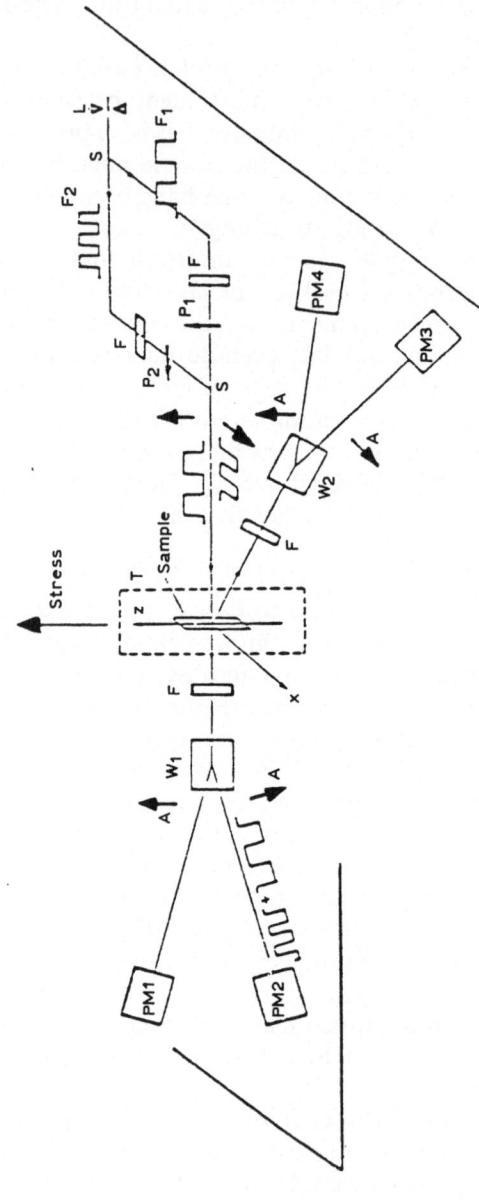

FIG. 5. Optical equipment: L, mercury lamp (HBO 200 W); S, beam splitter; $F_1$, $F_2$, modulation frequencies of the mechanical choppers; F, optical filter; $P_1$, $P_2$, polarizers; $W_1$, $W_2$, Wollaston prisms; A, analysing direction; PM, photo-multiplier; T, temperature chamber.

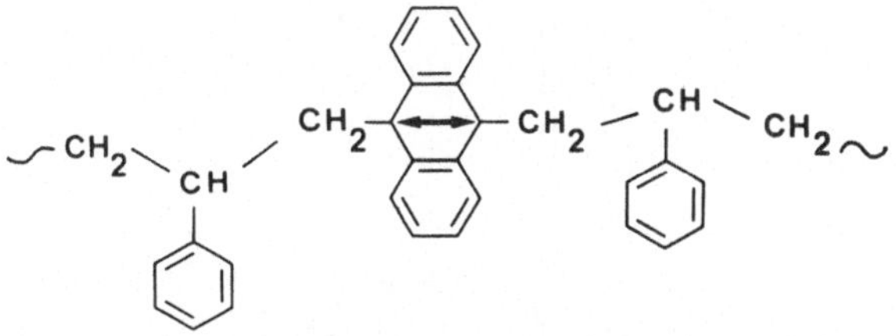

FIG. 6. Labelled chain containing an anthracene fluorescence group (the arrow represents the transition moment).

It is worth noting that the orientation functions $\langle P_2(\cos\theta)\rangle$ and $\langle P_4(\cos\theta)\rangle$ obtained from FP measurements on labelled chains reflect only the orientation of the nearest neighbour chain segments of the fluorescent molecule. The average deals with all the labelled chains in the sample.

## 5. ORIENTATION AND CHAIN RELAXATION OF POLYSTYRENE UNIAXIALLY STRETCHED ABOVE $T_g$

The techniques described above were first applied to the molecular behaviour of polystyrene chains. The sample stretching is performed on a specially designed machine,[16] shown in Fig. 7. This machine operates at constant stain rate, $\dot{\varepsilon}$ (in the range from $2 \times 10^{-3}\,s^{-1}$ to $2 \times 10^{-1}\,s^{-1}$), up to a 600% deformation for a sample of 6 cm length between the jaws. The stress is simultaneously recorded. The temperature controlled chamber can be regulated from room temperature up to 150°C. The temperature is controlled to within $\pm 0.002$°C and is homogeneous along the stretching axis to at least 0.033°C.

In the following, the draw ratio is defined as $\lambda = l/l_0$ ($l_0$ is the initial length of the sample between the jaws, $l$ is the length after drawing). The stress has to be understood as the nominal stress defined by $\sigma = F/S_0$ where $F$ is the tensile force and $S_0$ is the initial cross-section of the sample.

Orientation studies reported hereafter have been obtained recently in our laboratory on a series of polystyrene samples, the characteristics of which are reported in Table 1.

### 5.1. Infrared Dichroism Studies[17]
Thin films suitable for IR spectroscopy have been obtained by solution casting. The PS films were stretched at the chosen temperature up to a given $\lambda$ and then suddenly quenched at room temperature. IRD

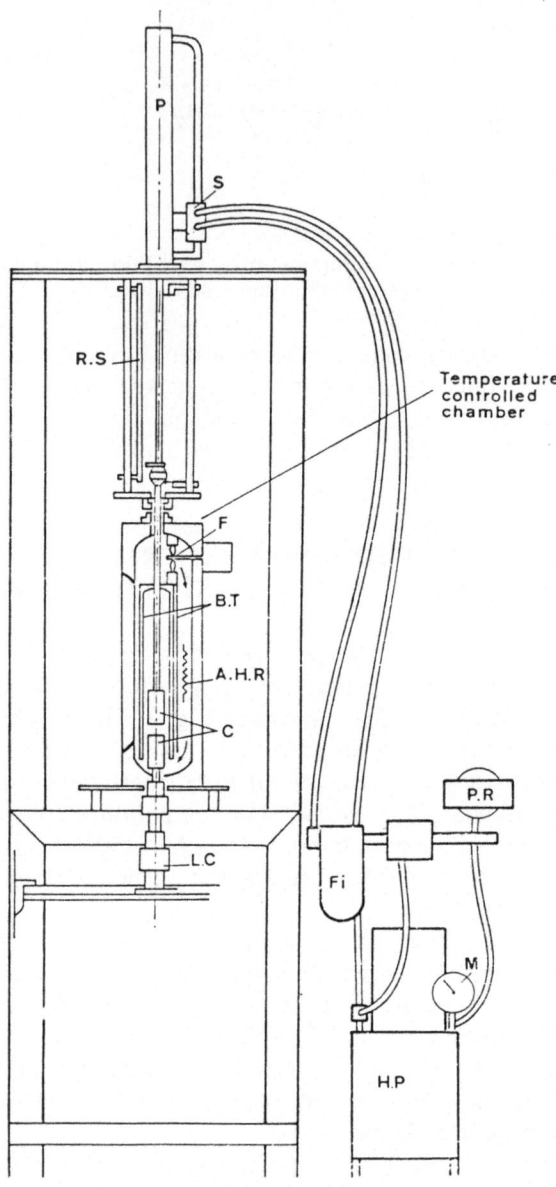

FIG. 7. Stretching machine and temperature controlled chamber: P, double hydraulic plunger; S, servovalve; RS, resisting sensor; C, clamps; PR, pressure reservoir; M, manometer; HP, hydraulic pump; LC, load cell; F, fan; BT, blowing tubes; AHR, additional heating resistances; Fi, filter.

## TABLE 1
### NUMBER-AVERAGE MOLECULAR WEIGHT AND POLYDISPERSITY OF POLYSTYRENE SAMPLES

| Polymer | $\bar{M}_n$ | $\bar{M}_w/M_n$ |
|---------|-------------|-----------------|
| PS 1 | 149 800 | 1·70 |
| PS 100 | 105 000 | 1·12 |
| PS 160 | 160 000 | 1·16 |
| PS 200 | 190 000 | 1·17 |
| PS 400 | 420 000 | 1·24 |
| PS 600 | 660 000 | 1·15 |
| PS 900 | 855 000 | 1·19 |
| PS 1 300 | 1 300 000 | 1·28 |

measurements were later performed using a Nicolet 7199 Fourier transform spectrometer.

The orientation refers to the local chain axis of PS shown in Fig. 2. The 1028 and 906 cm$^{-1}$ absorption bands used are not sensitive to chain conformation, and thus the orientation function $P_2(\cos\theta)$ is averaged over all segments of a chain and over all the chains in the sample; it will be given as $\langle P_2(\cos\theta)\rangle_{av}$.

Let us consider the polydisperse sample PS 1. Figure 8 illustrates the change in the orientation function $\langle P_2(\cos\theta)\rangle_{av}$ as a function of draw ratio $\lambda$ for a strain rate $\dot{\varepsilon} = 0·115\,\mathrm{s}^{-1}$ and different stretching temperatures. Up to $\lambda = 3·5$ this change approximates to a straight line, the slope of which decreases with increasing temperature.

Figure 9 shows the influence of strain rate $\dot{\varepsilon}$ for a stretching temperature of 122°C. Similarly, a linear relationship holds, for different strain rates, between $\langle P_2(\cos\theta)\rangle_{av}$ and draw ratio. For a given draw ratio, $\langle P_2(\cos\theta)\rangle_{av}$ increases with strain rate.

These results suggest that relaxation of orientation occurs during stretching. In order to get a more quantitative analysis of these relaxation phenomena, the results obtained by infrared dichroism were analysed according to the method developed by Lodge[18] for birefringence. The problem is to describe the orientation of a polymer during deformation by a constitutive equation, analogous to stress. Among the different constitutive equations that have been proposed to describe the viscoelastic properties of polymers, we chose the equation corresponding to the Lodge model. In this model, an amorphous polymer is considered during

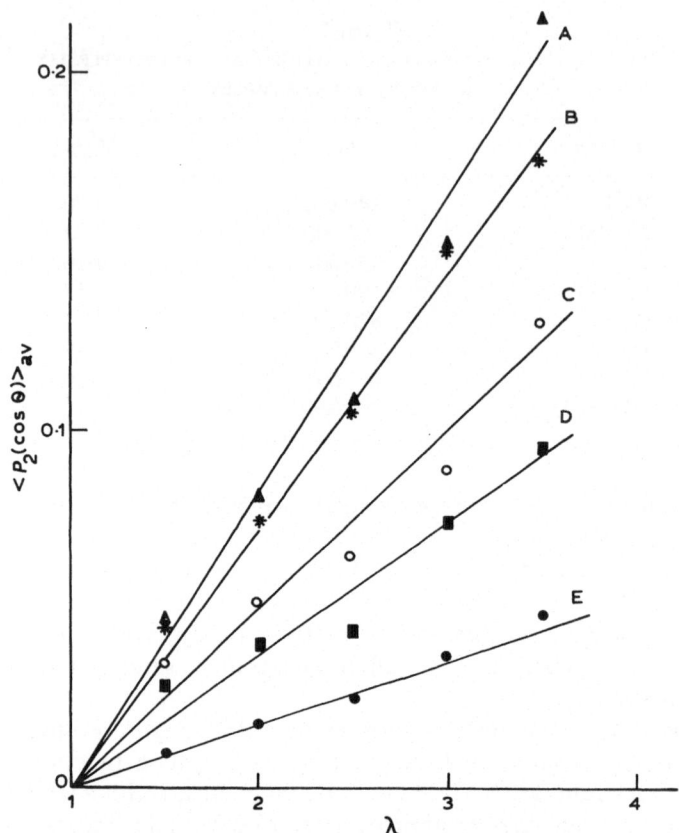

FIG. 8. Orientation function of PS 1 as a function of draw ratio: strain rate $\dot{\varepsilon} = 0.115\,\mathrm{s}^{-1}$; temperature of stretching; A, 110°C; B, 113°C; C, 116·5°C; D, 122°C; E, 128·5°C.

deformation as a network of elastic chains, the density of links of which changes with time.

As far as stretching at constant strain rate is concerned, the second-order moment of the orientation function $\langle P_2(\cos\theta)\rangle_{av}$ will be given by the constitutive equation

$$\langle P_2(\cos\theta)\rangle_{av}$$
$$= \sum_i \theta_i \left[ \frac{3\dot{\varepsilon}\tau_i}{(1 - 2\dot{\varepsilon}\tau_i)(1 + \dot{\varepsilon}\tau_i)} - \frac{2\dot{\varepsilon}\tau_i \exp(2\dot{\varepsilon}t - t/\tau_i)}{1 - 2\dot{\varepsilon}\tau_i} - \frac{\dot{\varepsilon}\tau_i - \exp(-\dot{\varepsilon}t - t/\tau_i)}{1 + \dot{\varepsilon}\tau_i} \right]$$
$$(20)$$

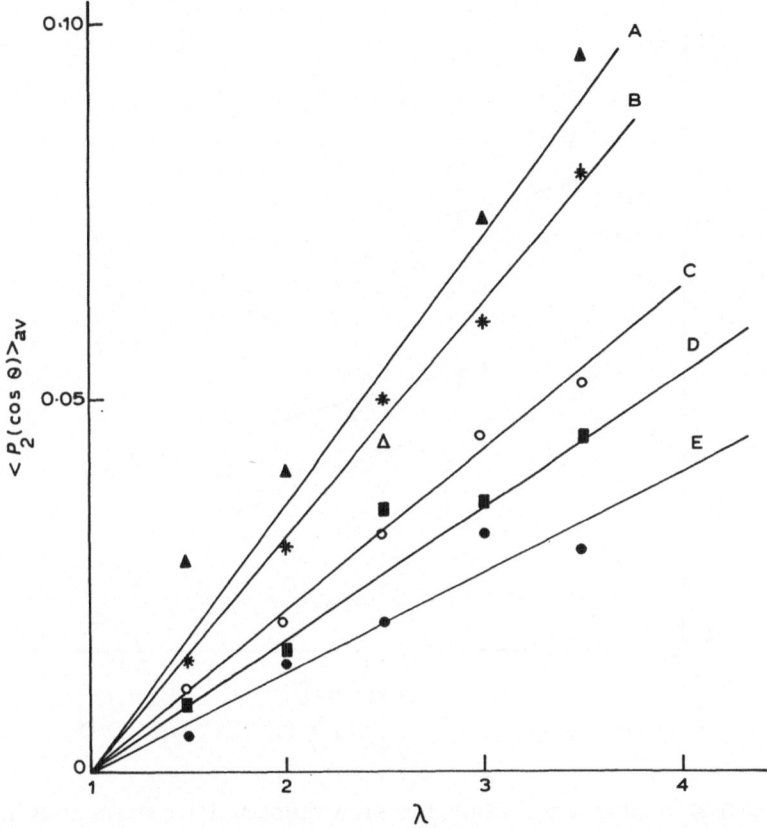

FIG. 9. Orientation function of PS 1 as a function of draw ratio: temperature of stretching $T = 122°C$; strain rate: A, $0.115 \, s^{-1}$; B, $0.086 \, s^{-1}$; C, $0.059 \, s^{-1}$; D, $0.026 \, s^{-1}$; E, $0.008 \, s^{-1}$.

where $\theta_i$ and $\tau_i$ originate from a decomposition of the orientation relaxation function $\theta(t)$ into a sum of exponentials:

$$\theta(t) = \sum_i \theta_i \exp\left(-t/\tau_i\right) \tag{21}$$

In order to perform the calculation we arbitrarily chose the relaxation function as

$$\theta(t) = \sum_{i=-1}^{+3} \theta_i \exp\left(-t/10^i\right) \tag{22}$$

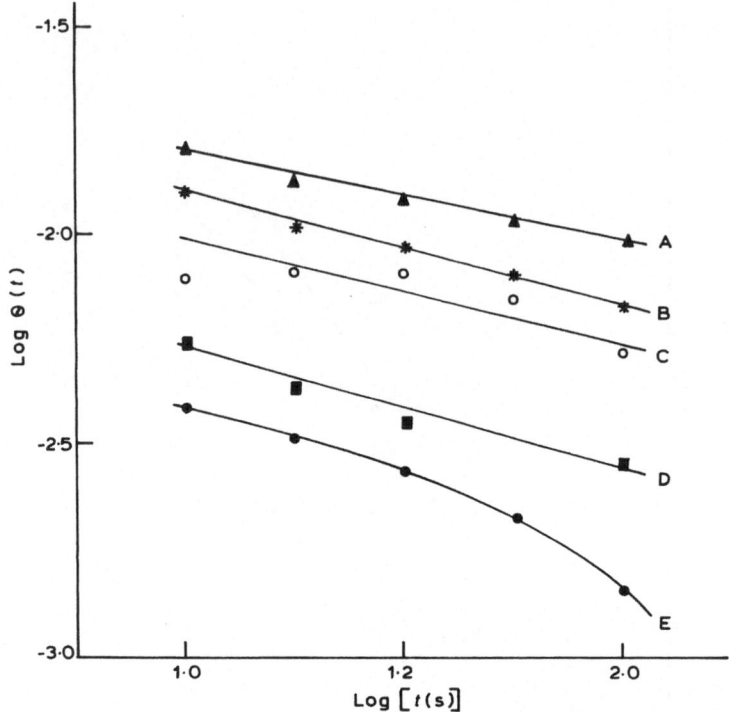

FIG. 10.  Log $\theta(t)$ vs. log $t$ for PS 1: A, 110°C; B, 113°C; C, 116·5°C; D, 122°C; E, 128·5°C.

For each stretching temperature, five draw ratios and five strain rates have been used to calculate the $\theta_i$. The change of log $\theta_i$ as a function of log $t$ for the five temperatures used is shown in Fig. 10. From these results, it is possible to get a master curve for the orientation relaxation function; curves from Fig. 10 shifted using the WLF coefficients previously obtained by Plazeck[19]

$$\log a_T = -\frac{10\cdot04(T-115)}{62\cdot6+(T-115)} \tag{23}$$

gave the curve shown in Fig. 11 for a reference temperature of 115°C.

As $\langle P_2(\cos\theta)\rangle_{av}$ seems to be a linear function of $\lambda$ in the range of accuracy of the measurements, another way to illustrate the orientation behaviour is to plot the value of $\langle P_2(\cos\theta)\rangle_{av}$ at a given $\lambda$ value as a function of $T$ and $\log(1/\dot\varepsilon)$. The choice of $\log(1/\dot\varepsilon)$ as a variable was suggested by the fact that in our experiments, realized at constant strain

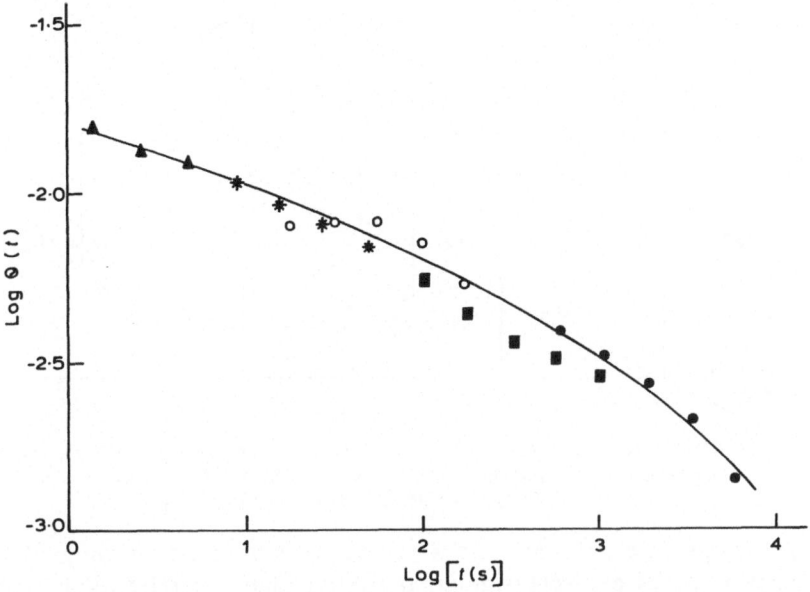

FIG. 11.   Master curve $\log \theta(t)$ vs. $\log t$ for PS 1 between 110°C and 128·5°C; reference temperature $T_0 = 115$°C.

rate, $1/\dot{\varepsilon}$ is similar to time. The influence of PS molecular weight on the change of $\langle P_2(\cos \theta)\rangle_{av}$ at $\lambda = 4$ as a function of temperature and strain rate is illustrated in Fig. 12. For a temperature of stretching close to the glass transition temperature, the two polymers behave in the same way. An increase of temperature results in a greater level of orientation for the higher molecular weight. The corresponding orientation relaxation functions are given in Fig. 13. Under the experimental conditions used, the two samples behave in the same way for short relaxation times. For longer relaxation times, the orientation relaxation function $\theta(t)$ decreases faster for PS 100. Such behaviour is clearly explained by the extent of the plateau region at long time with increase in molecular weight, as found from the mechanical viscoelasticity studies.

It should be noted that the behaviour of the polydisperse sample (PS 1), found to be similar to PS 100, points to the great influence of the low molecular weight chains on the orientation and relaxation of polydisperse PS.

In order to compare our results with the relaxation modulus $E(t)$, viscoelasticity measurements have been performed using dynamic shear

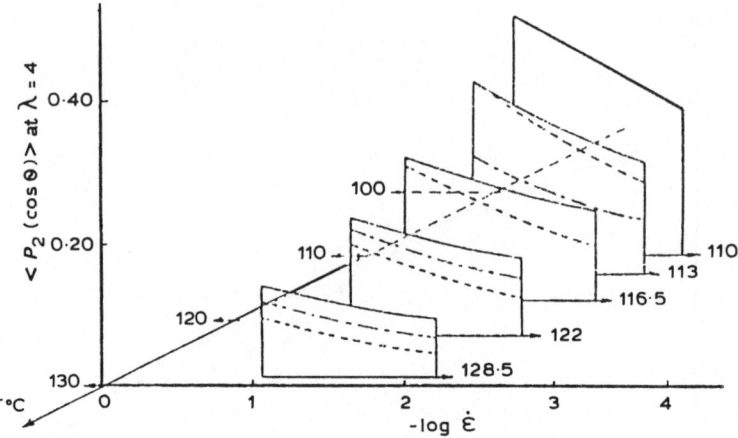

FIG. 12.   Orientation function at $\lambda = 4$ as a function of temperature and strain rate for monodisperse polystyrene samples: (———) PS 900 from IRD; (– – – –) PS 100 from IRD; (– · –) PS 400–PAP 370 from FP.

experiments. In Fig. 14, $E(t)$ and $\theta(t)$ curves are compared for sample PS 1. The two curves are very similar, indicating that the relaxation of the orientation is closely related to the plateau region and the beginning of the terminal zone.

## 5.2.  Fluorescence Polarization Studies
The fluorescence polarization measurements are carried out during

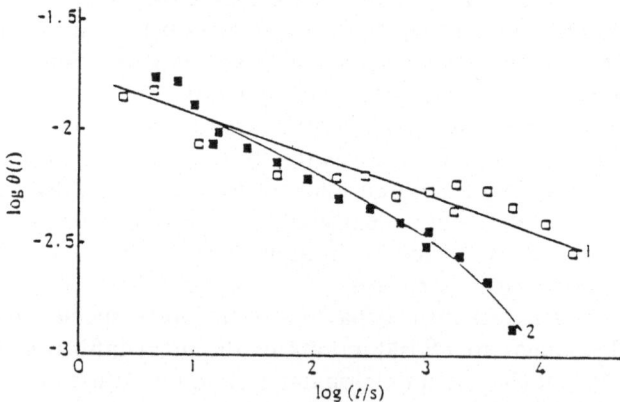

FIG. 13.   Master curves of $\log \theta(t)$ plotted against $\log t$ for monodisperse samples between 110°C and 128·5°C; reference temperature $T_0 = 115$°C. (1) PS 900; (2) PS 100.

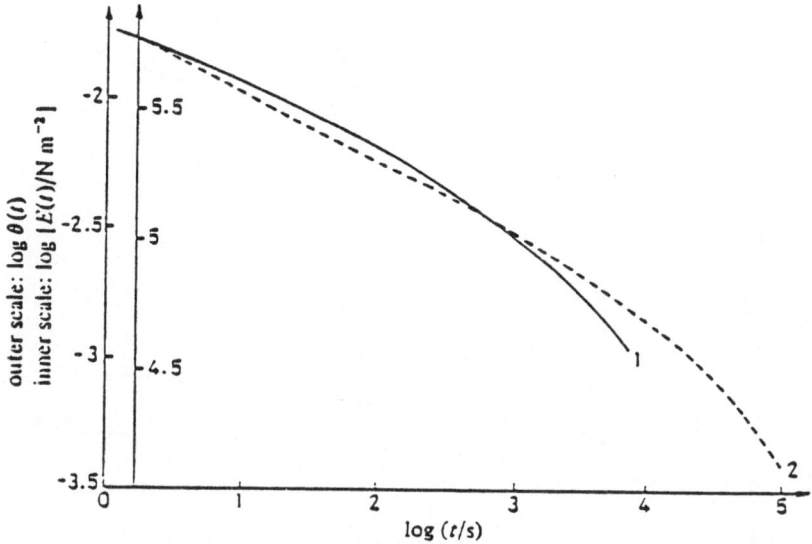

FIG. 14. Comparison between $\log \theta(t)$ and $\log E(t)$ plotted as a function of $\log t$ for PS 1; reference temperature $T_0 = 115°C$. (1) IRD orientation relaxation; (2) mechanical relaxation.

stretching, using the optical equipment shown in Fig. 5 adapted to the stretching machine shown in Fig. 7.

The samples are moulded under pressure and annealed to obtain transparent bubble-free samples and to avoid any birefringence. Sample size is 8 cm length, 2 cm width and 0·2 cm thickness.

### 5.2.1. Effects of Temperature and Strain Rate[16]

Anthracene labelled polystyrene chains (called PAP) with a number average molecular weight, $\bar{M}_n = 287\,000$ have been incorporated (at 1% w/w concentration) in a polystyrene matrix (PS 200) with a narrow molecular weight distribution ($\bar{M}_n = 191\,000$, $\bar{M}_w = 207\,000$). $T_g$ of the resultant polymer blend, measured by differential scanning calorimetry at a heating rate of $10°C\,min^{-1}$ is 107·5°C.

A typical example of stress–strain and orientation–strain curves is given in Fig. 15. In this temperature range, polystyrene samples undergo a homogeneous deformation. The stress–strain curves show two main regions. The first, at small extension ratio, is characterized by a rapid increase of stress and is attributed to the glassy deformation. The second, at higher extension ratio, in which the stress increases slowly with the

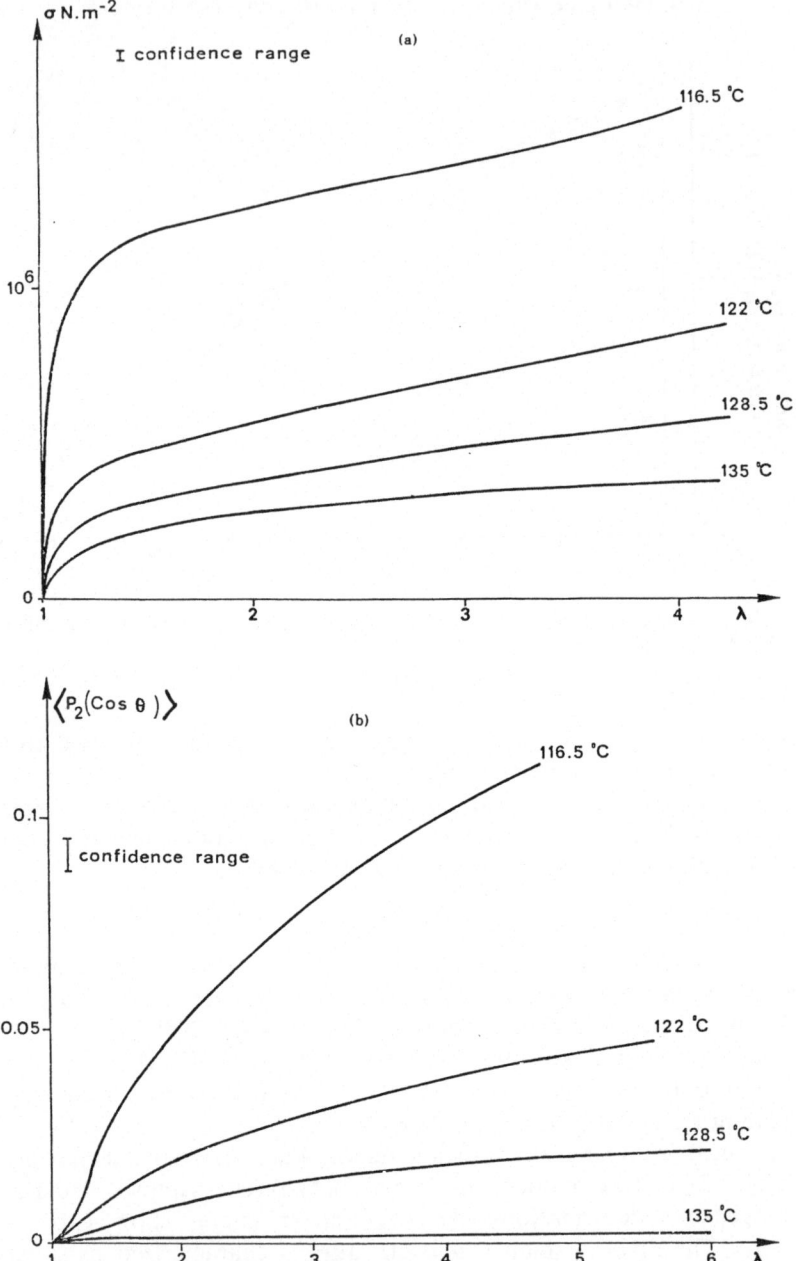

FIG. 15. (a) Stress and (b) orientation function vs. draw ratio as a function of temperature; strain rate, $\dot{\varepsilon} = 0{\cdot}115\,\mathrm{s}^{-1}$.

deformation, reflects a rubber-like deformation, with a possible contribution from flow. In contrast, orientation curves, in which the orientation is plotted against the extension ratio $\lambda$, show a regular increase of orientation during stretching.

The influence of temperature appears clearly on these two plots. It can be seen that, at a given strain rate, the first part of the stress–strain curve is very sensitive to temperature whereas the slope of the second part decreases as temperature rises. It can also be noted that the orientation falls continuously and at 135°C is too low to be measurable.

Figure 16 shows the behaviour of stress and orientation at $T_g + 9°C$ for various strain rates. It is clearly seen that only the first part of the stress–strain curve is affected by the strain rate, but the orientation is not modified within the accuracy of the experiments. Comparison with Fig. 17 shows that, at a higher temperature of $T_g + 14·5°C$, both the stress reflecting the rubber-like behaviour and the orientation are functions of the strain rate.

It must be noted that the orientation given by the fluorescence polarization measurements refers to the anthracene group, the transition moment of which lies along the chain axis and is located in the middle of the chain. This particular position implies that the measured orientation reflects the orientational behaviour of statistical chain segments of the central part of the chain. It seems difficult to state the size of this central part precisely. Owing to the molecular weight of the labelled chain ($M_n = 287\,000$), it is reasonable to assume that the orientation of the chain-end segments is not taken into account by this method.

Comparison between Figs 15(a) and 15(b) seems to indicate that orientation does not originate from the glassy part of the stress but is related to the rubber-like component. At a given temperature, the slope of the rubbery part of the stress–strain curve and its dependence on the strain rate govern the orientation behaviour. The shrinkage experiments performed on deformed samples at a temperature of 120°C for 24 h show that, ' within the experimental conditions (strain rate, temperature, molecular weight), the contribution of the terminal flow zone is negligible.

Thus, in these experiments, the orientation observed through fluorescence polarization is not directly related to the complete true stress, but depends on the deformation in the rubbery plateau zone. This result is rather different from that which is obtained with birefringence measurements in which stress and orientation are related through a constant coefficient. Also the IRD measurements reported above with films produced from the same polystyrene batch lead to the conclusion

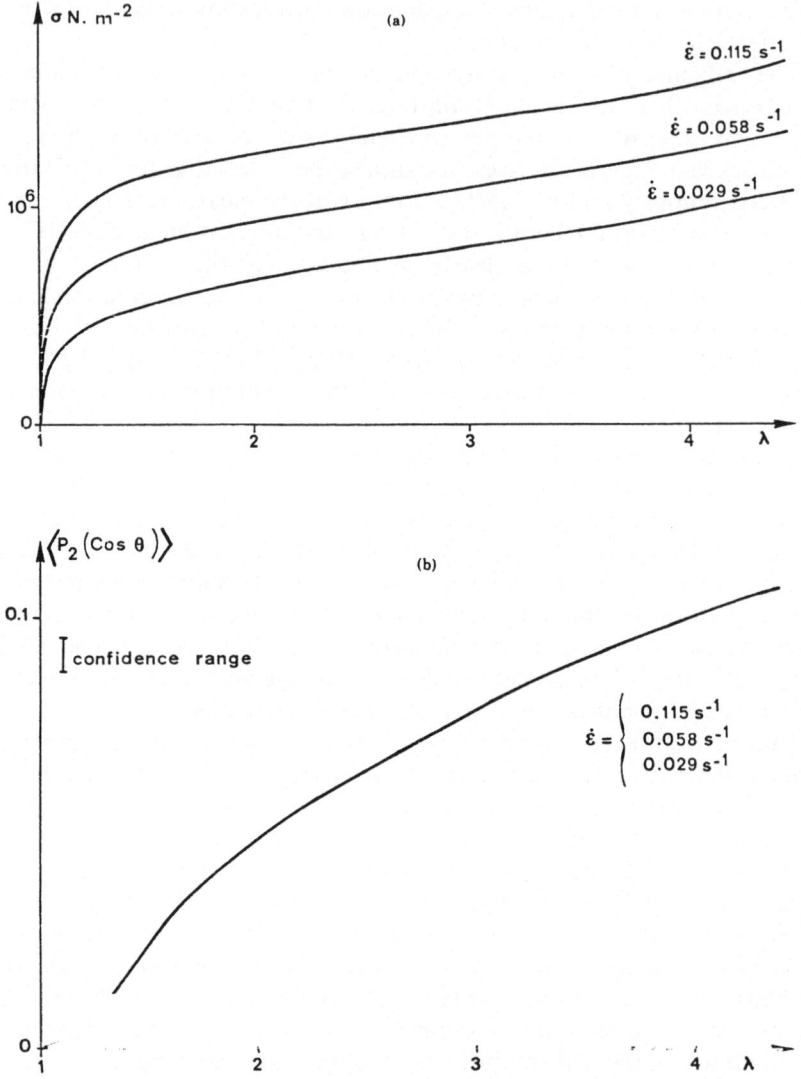

FIG. 16.   (a) Stress and (b) orientation function vs. draw ratio as a function of
strain rate; stretching temperature, $T = 116.5°C$.

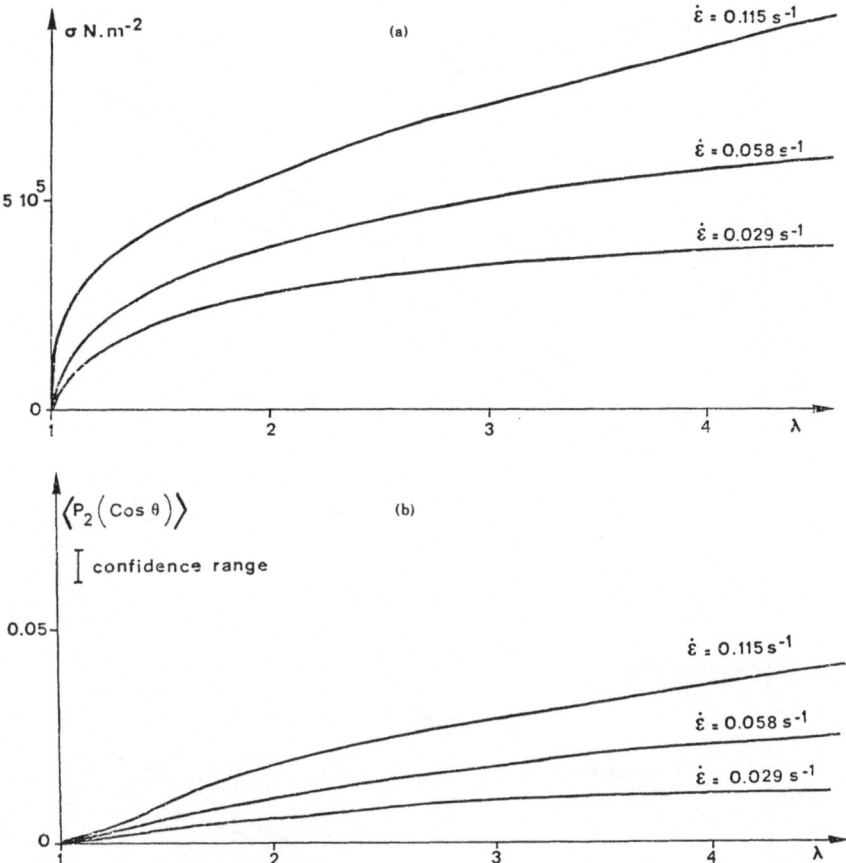

FIG. 17.   (a) Stress and (b) orientation function vs. draw ratio as a function of strain rate; stretching temperature, $T = 122°C$.

that stress and orientation obtained by IR dichroism are identical. However, it is worth noting that, in the rubbery plateau zone, infrared dichroism and fluorescence polarization lead to similar behaviour with respect to temperature and strain rate dependences, as illustrated in Fig. 12 where $\langle P_2(\cos\theta)\rangle$ values at $\lambda = 4$ obtained by FP measurements on labelled polystyrene chains PAP 370 with a number average molecular weight, $\bar{M}_n = 370\,000$, incorporated in a PS 400 matrix, are compared to the values determined by IRD for PS 100 and PS 900. At 110°C and 120°C, the behaviour observed by FP for PS 400 is intermediate between the IRD behaviours of PS 100 and PS 900, as expected. The discrepancy

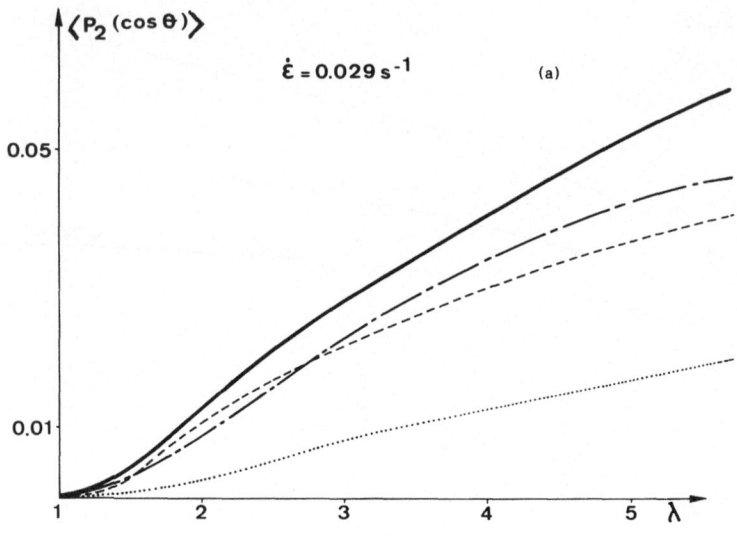

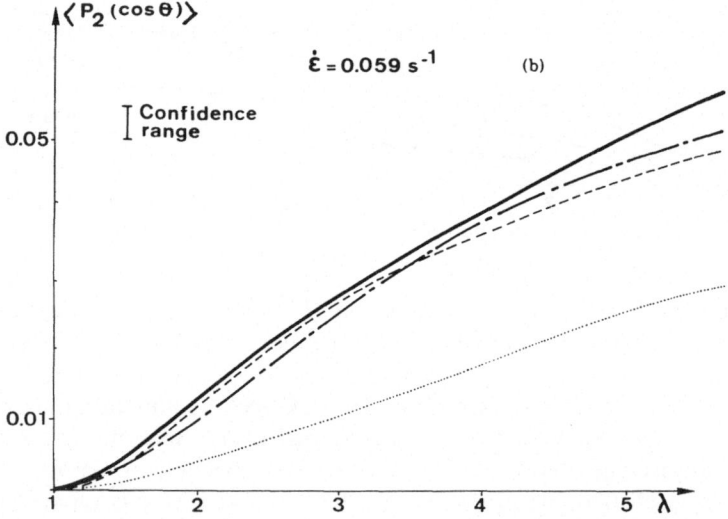

FIG. 18. Orientation function vs. draw ratio at various strain rates: (a) $\dot{\varepsilon} = 0.029\,s^{-1}$; (b) $\dot{\varepsilon} = 0.059\,s^{-1}$; (c) $\dot{\varepsilon} = 0.115\,s^{-1}$. The curves correspond to ($\cdots$) PS 200, (---) PS 400, (-·-) PS 600, (——) PS 900 and PS 1300.

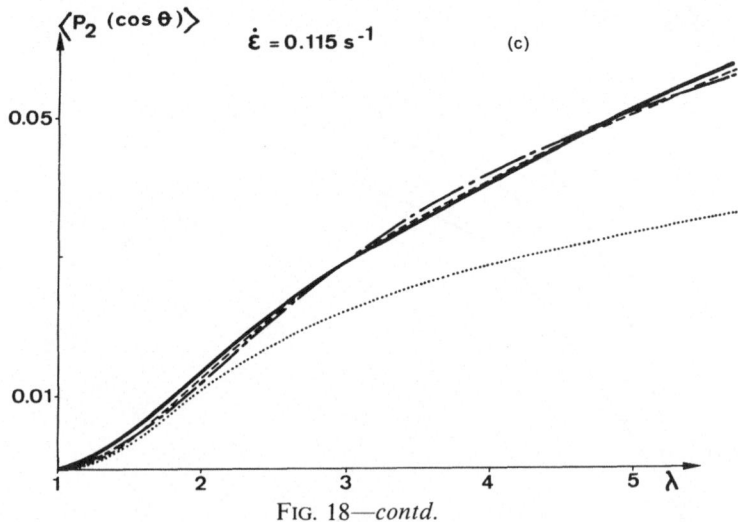

FIG. 18—contd.

between IRD and FP behaviour observed at lower temperatures, where the glassy part of the stress has a large contribution at small strains, will be discussed in Section 5.2.3.

From results presented in Figs 15(b) and 17(b) it can also be noted that an increase in temperature may have the same effect on orientation as a decrease in strain rate, so time and temperature seem to affect the system in the same way. Such behaviour must be attributed on a molecular level to the participation of relaxation phenomena during stretching.

### 5.2.2. Influence of the Molecular Weight of the Polymer Matrix

In order to go further into the investigation of chain relaxation during stretching, a set of experiments[20] has been performed on the influence of the molecular weight of the polystyrene matrix on the orientation of the labelled polystyrene PAP 287 ($\bar{M}_n = 287\,000$). The stretching was performed at 128·5°C for various constant strain rates $\dot{\varepsilon}$. The molecular weight characteristics of the matrix chains are reported in Table 1.

The orientation function of the PAP chain in the various matrices is shown as a function of the draw ratio in Fig. 18. It can be seen from these plots that the orientation of the PAP chains is the same in PS 900 and PS 1300 matrices at a given strain rate, and quantitative comparison of these curves leads to the conclusion that the orientation of the labelled chain is independent of the strain rate for these two matrices. This is not the case for the other matrices. For $\dot{\varepsilon} = 0.029$ or $0.059\,\text{s}^{-1}$ the orientation of the

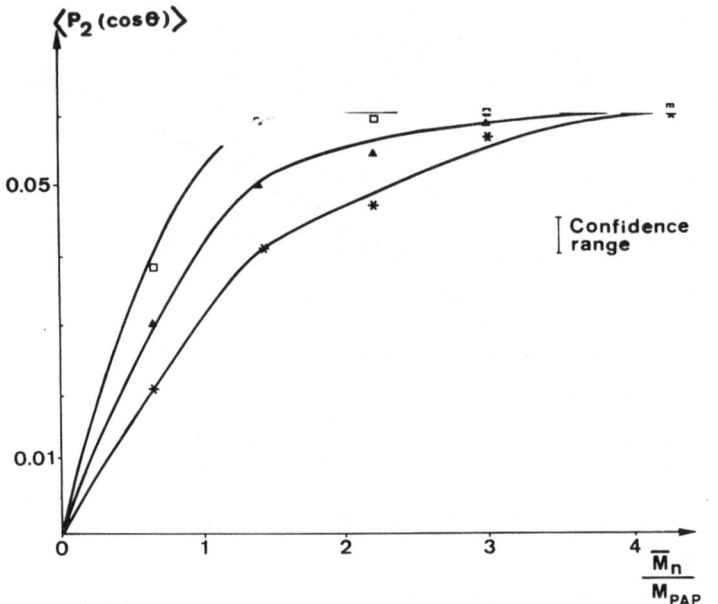

FIG. 19.   Orientation function, measured at draw ratio $\lambda_0 = 6$, as a function of the strain rate vs. the ratio $M_{matrix}/M_{PAP}$: ($\square$) $\dot{\varepsilon} = 0.115\,s^{-1}$; ($\blacktriangle$) $\dot{\varepsilon} = 0.059\,s^{-1}$; ($*$) $\dot{\varepsilon} = 0.029\,s^{-1}$.

PAP decreases as the molecular weight of the matrix decreases. This effect tends to disappear when stretching is performed at higher strain rate (i.e. $0.115\,s^{-1}$) where the behaviour of the fluorescent chain is the same within the accuracy of our experiments in the matrices from PS 400 to PS 1300, but the orientation is still significantly lower in PS 200 ($M_{matrix} < M_{PAP}$).

The influence of the strain rate and molecular weight of the matrix is better demonstrated in Fig. 19. The orientation function $\langle P_2(\cos\theta)\rangle$ measured at a draw ratio $\lambda_0 = 6$ is plotted against the ratio of the molecular weights of the matrix and the labelled chains for different strain rates. This variable affords a comparison between the relative lengths of the PAP chains and the matrix chains.

At this point, it is of interest to note that the measured orientation is related to the behaviour of the central part of the chain and does not basically represent an average over all chain units. During stretching, polymer chains may experience relaxation. The characteristic time of an experiment at a given strain rate $\dot{\varepsilon}$ is on the order of $\dot{\varepsilon}^{-1}$, so only relaxation mechanisms with times in the range of $\dot{\varepsilon}^{-1}$ can contribute.

From this point of view, some features appearing in Fig. 19 can be explained qualitatively. If the length of the chains of the matrix is greater than that of the PAP chains, at a given $\dot{\varepsilon}$ the relaxation of the labelled chain becomes slower, leading to higher orientation. On the other hand, if the surrounding chains are shorter than the PAP chains, the relaxation of these species is more important and lower orientation is observed. Such an experimental fact suggests a strong coupling between the relaxation of the labelled chain and the relaxation of the matrix chains. The first macroscopic effect of this interaction consists in an apparent modification of the relaxation times of the labelled chain depending on the length of the surrounding chains. These conclusions are valid for the strain rates used here.

When the molecular weight of the matrix increases, the apparent relaxation times are shifted toward larger values in such a way that no further influence of the strain rate can be discerned, under the experimental conditions considered here, and $\langle P_2(\cos\theta)\rangle$ approaches a limiting value. When the strain rate is lower, the time domain involved is larger and a higher matrix molecular weight is required to reach this limit of the orientation function.

A deeper insight can be reached by considering the molecular models of polymer viscoelasticity which have recently been developed by Doi and Edwards[2,3] based on the tube model. In this model, a chain is constrained within a hypothetical tube representing the highly entangled surrounding chains. Such a tube is fixed relative to the motion of the individual chain. The tube constraints have been modelled by Doi and Edwards through the slip-link model, in which a chain is trapped by small frictionless rings, through which it can pass freely and which represent the topological constraints. The spatial position of the slip-links is also fixed relative to the motion of the chain.

The relaxation of a suddenly deformed polymer melt, the strain then being kept constant, is described in terms of three different relaxation processes occurring on various time scales.

We start from an affinely deformed chain and consider the relaxation of this chain in fixed surroundings. At short times, the slip-links act as fixed crosslinks and the chain can relax between these entanglement points like a Rouse chain with fixed ends. This first relaxation time $\tau_A$, is independent of the length of the chain, and only depends on the average number of monomers between entanglement points, $N_e$. This relaxation process is quite fast at a temperature above the glass transition temperature and it will not be considered later.

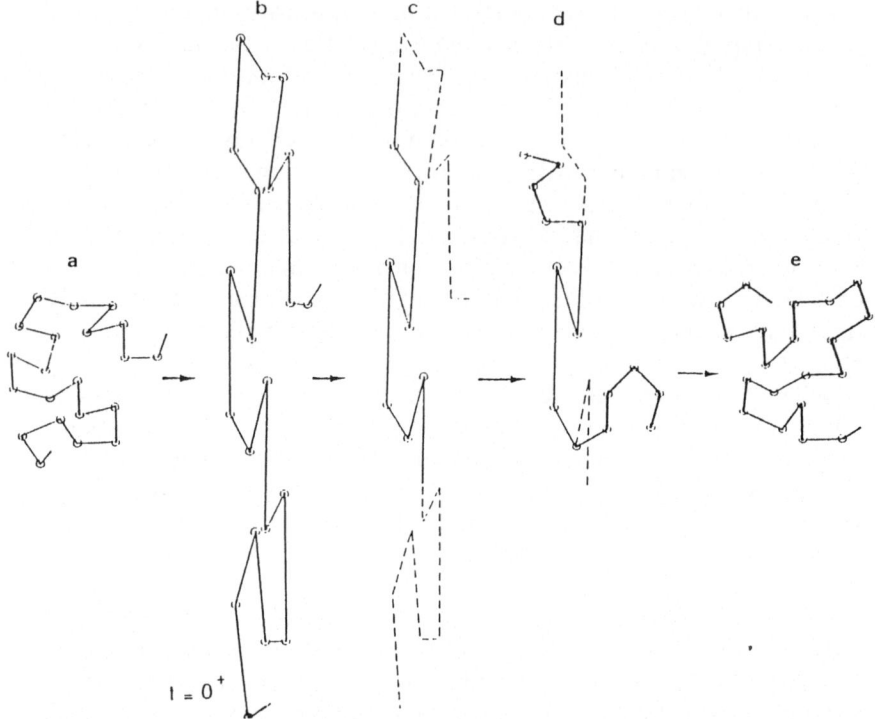

FIG. 20.   A given chain embedded in a network of chemically identical chains. Different relaxation stages of the primitive chain after a step uniaxial stretching: (a) initial isotropic state; (b) step-strained primitive chain at time $0^+$; (c) primitive chain at the end of the self-retraction process; (d) primitive chain during reptation; (e) primitive chain after complete disengagement. Slip-links are represented by small circles. For easier comparison, in each figure the state of the primitive chain at the end of the previous stage is represented by a dashed line. Newly created parts of the tube are drawn in heavy line.

During the second relaxation process, which is shown schematically in Fig. 20, the chain shrinks into the deformed tube in order to equilibrate with respect to the stretch along itself and to recover the equilibrium monomer density per unit arc length. This process is characterized by a relaxation time $\tau_B$, related to $\tau_A$ by

$$\tau_B = 2(N_0/N_e)^2 \tau_A \tag{24}$$

where $N_0$ is the number of monomers per chain.

In the last stage, shown in Fig. 20, the chain disengages itself from the original deformed tube by the reptation process proposed by de Gennes.[1]

This occurs in a time $\tau_C$ such that

$$\tau_C = 6(N_0/N_e)^3 \qquad \tau_A = 3(N_0/N_e)\tau_B \qquad (25)$$

Analytical expressions for these three relaxation times show that if $N_0 \gg N_e$ (which is a basic condition for these models of entangled chains) the processes are well separated in time.

Returning to the results obtained at a given strain rate, according to this model the same orientation of the labelled chain should be found independently of the molecular weight of the matrix when $M_{matrix} > M_{PAP}$, i.e. if the environment is fixed relative to the motion of the labelled chain. Such behaviour is observed qualitatively at the highest strain rate $(0.115 s^{-1})$ but for lower rates the orientation still increases with increasing molecular weight of the matrix. Before trying to improve this model, it is useful to determine the relaxation motions involved in these experiments. The first relaxation process occurs at very short times $\tau_A$ and does not create a change in the orientation since the distribution of the end-to-end vector of the labelled chains is not affected, nor is any vector joining two consecutive entanglement points. With regard to the third relaxation process, the fact that the deformation was recoverable by annealing above $T_g$ proves that no flow occurs and thus the reptation time $\tau_C$ was not reached under the experimental time and temperature conditions. Hence, in these experiments the orientation is mostly affected by the shrinking of the chain into its deformed tube.

During stretching, both PAP chains and matrix chains experience retraction occurring respectively in times $\tau_B^{PAP}$ and $\tau_B^{matrix}$. As in these experiments the labelled chain is the same in the different matrices, the retraction process can be checked by plotting the orientation at a given draw ratio as a function of the dimensionless parameter $\dot{\varepsilon}\tau_B^{matrix}$, since in a stretching experiment we are dealing with a time range of the order of $\dot{\varepsilon}^{-1}$. In fact, as $\tau_B^{matrix}$ scales as $M^2$, it is equivalent to plot the orientation function versus the quantity $\dot{\varepsilon}M_{matrix}^2$. As shown in Fig. 21, all the points fall fairly well on a single curve, proving that the orientation function under our experimental conditions is completely controlled by the chain retraction process.

Such an effect of the molecular weight of the polymer matrix on the retraction time of a labelled chain is not predicted for the long chains considered in this study. Indeed, if we consider that $N_e$ for polystyrene is around 160, the labelled chain participates to about 20 entanglements and even the smallest matrix molecular weight corresponds to more than 10 entanglements per chain. In such conditions, the Doi–Edwards theory

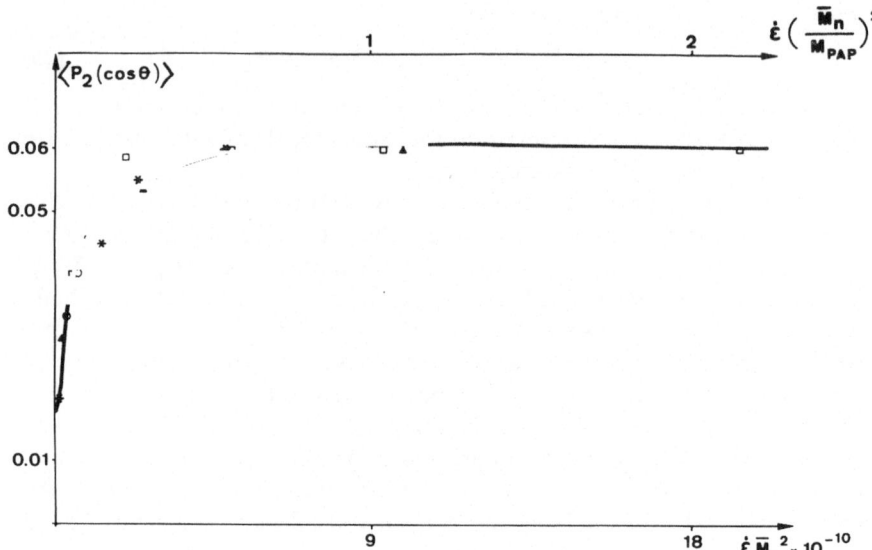

FIG. 21. Orientation function, measured at $\lambda_0 = 6$, vs. the parameter $\dot{\varepsilon}M_{matrix}^2$: data obtained from ($\square$) $\dot{\varepsilon} = 0.115\,\mathrm{s}^{-1}$; ($\blacktriangle$) $\dot{\varepsilon} = 0.059\,\mathrm{s}^{-1}$; ($*$) $\dot{\varepsilon} = 0.029\,\mathrm{s}^{-1}$; ($\bigcirc$) data corresponding to $M_{matrix} = M_{PAP}$.

does not predict any coupling between the behaviour of one given chain and that of the surrounding chains. The Doi–Edwards model would correspond to a free chain in a polymer network for the surrounding chains are not considered as able to retract or reptate. For matrix chains of a finite length, after a step deformation, the retraction process inside their own tubes and the reptation process out of the original tubes occur. During these relaxation processes, the constraints on the labelled chain arising from the matrix chains will be changed. For example, when a matrix chain end reptates away from the neighbourhood of the labelled chain, this latter has an opportunity to escape out of its original tube. During this time another matrix chain (or the same one) may reptate across the old tube so that the tube of the labelled chain is permanently modified. In this way, the labelled chain will relax faster. This coupling mechanism between the relaxation of a labelled chain and the reptation of the surrounding chains is called tube renewal and it occurs both for deformed and undeformed chains. The same type of mechanism has been applied to deformed chains considering the retraction processes.[21] The effect of the retraction processes of the surrounding chains on the

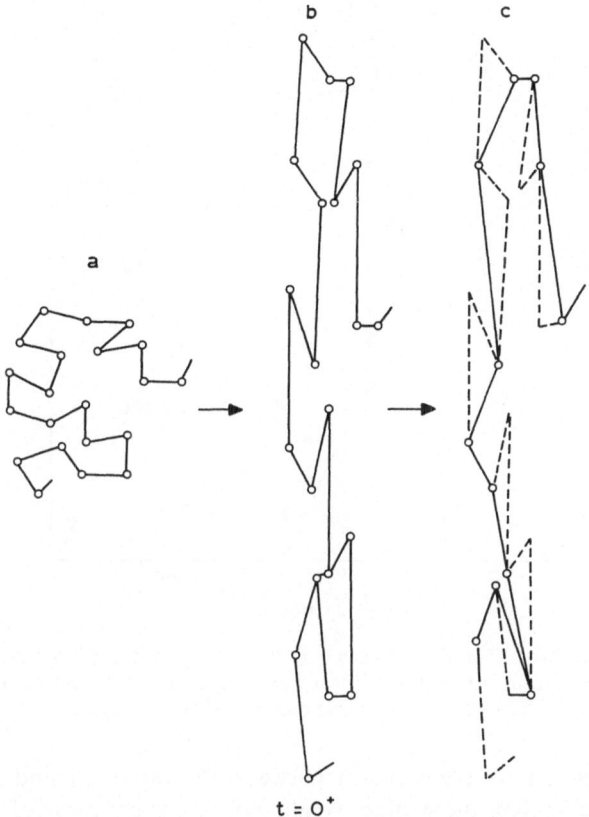

FIG. 22.   Modification of the path of a labelled chain due to the retraction of the matrix chains: (a) isotropic initial stage; (b) step-stretched primitive chain of the labelled chain at time $0^+$; (c) path of the primitive chain of the labelled chain after retraction of the surrounding chains, showing the local loosening due to random loss of slip-links.

retraction of the labelled chain is shown schematically in Fig. 22. As the retraction mechanism occurs only in deformed systems, the coupling shown in Fig. 22 and called tube relaxation is specific to these systems. Of course, the relaxation of the labelled chain is enhanced by the tube relaxation mechanism, and the shorter the matrix chains the faster the labelled chain relaxes. As the retraction characteristic time scales as $M^2$, the effect of the matrix molecular weight on the contribution of the tube relaxation to the relaxation of the labelled chain should be proportional to $M^2_{matrix}$. This matrix molecular weight dependence is actually observed, as

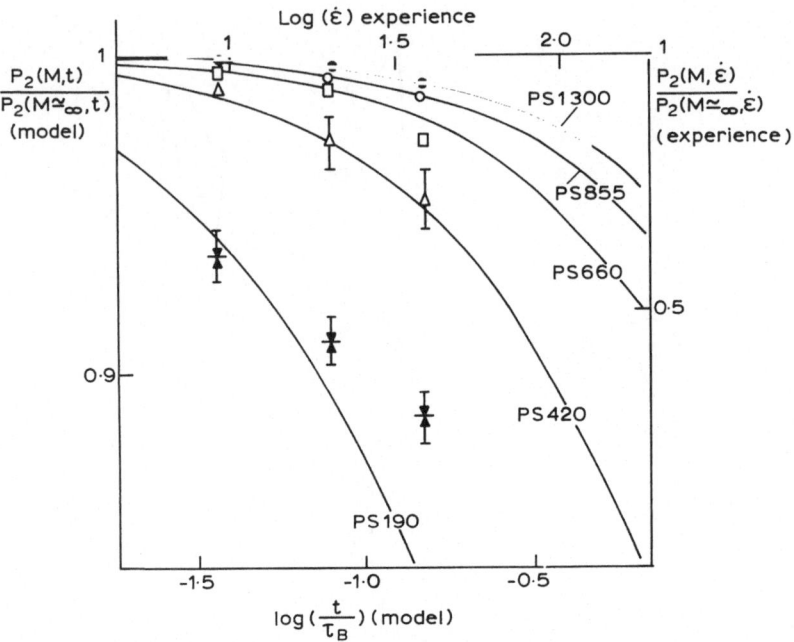

FIG. 23. Comparison of the observed and calculated matrix molecular weight effect on the orientation of an anthracene-labelled polystyrene chain. The full curves represent the calculated behaviour.

shown in Fig. 21. A comparison between the observed and calculated effects of the matrix molecular weight on the orientation of a labelled chain is shown in Fig. 23.

Satisfactory agreement is obtained, proving the validity of the proposed improvement of the Doi–Edwards model by this self-consistent approach.

### 5.2.3. Influence of the Molecular Weight of Labelled Chains

Another interesting feature is the influence of the molecular weight of the labelled chain when that of the polymer matrix is kept constant. Thus experiments have been recently performed in our laboratory[22] at 128·5°C with a PS 160 matrix. Results are presented in Fig. 24 at a strain-rate value $\dot{\varepsilon} = 0\cdot115\,s^{-1}$; similar behaviour is observed at other strain rates.

First, there is an increase in orientation with the molecular weight of the labelled chain. However, a more surprising result is the rather high orientation obtained for PAP 17 chains. Indeed, as the molecular weight of this labelled polymer is around the mean molecular weight between

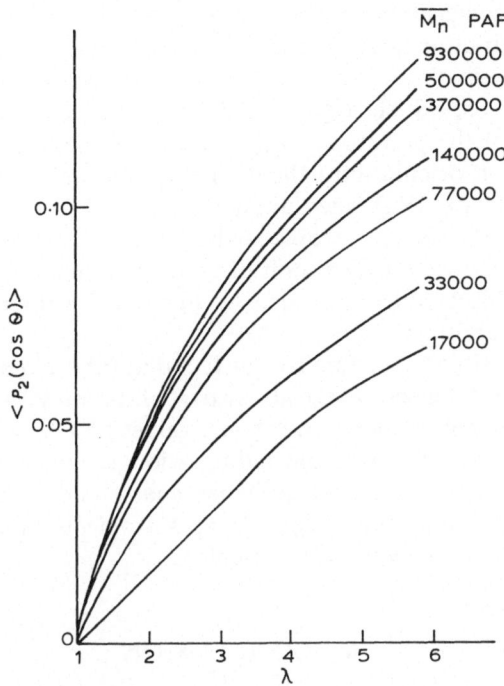

Fig. 24.   Orientation function plotted against draw ratio for various molecular weights of the PAP-labelled chains in a PS 160 matrix; stretching temperature, $T = 128 \cdot 5°C$.

entanglements for PS ($M_e \simeq 16\,000$), one would have expected a rather low orientation or even no orientation if only topological constraints were considered. Such chains are too short to be oriented efficiently by the deformation of the physical network. On the other hand, it seems that their orientation comes from the anisotropy of the surrounding medium. Note that similar orientation effects have been observed for pendant polyisoprene chains in a chemically crosslinked network.[23] The pendant chains labelled inside the chain or at the end of the chain exhibit an orientation which increases with $\lambda$, although always remaining lower than the orientation of the labelled chain involved in the permanent network. Furthermore, free fluorescent probes, made up of 9,10-dialkylanthracene with $16\,CH_2$ groups, are oriented at $\lambda > 3$. Thus from these results it appears that, in addition to orientation arising from topological effects (a physical entanglement network) and which could be described by molecular-viscoelasticity theories based on the tube concept, there is another contribution arising from the interaction with the surrounding anisotropic medium.

Although it could be argued that this effect is specific to FP and originates from the perturbation introduced by the fluorescent label, the fact that the FP orientation behaviour observed at 128·5°C is similar to that found from IRD measurements (Fig. 12) indicates that the contribution from the anisotropic medium is also involved to some extent in IRD. Indeed, the statistical unit of a polymer chain corresponds to an anisotropic object which can also be oriented by interaction with the strained surroundings.

It is now possible to account for the fact that FP orientation, unlike that of IRD, does not depend on the glassy part of the stress, which means that FP orientation only appears over some values of $\lambda$ ($\lambda > 1·2-1·3$). Indeed, the label inside the chain should induce a longer anisotropic object than the statistical segment of the polymer chain, and it has been shown experimentally[24] that the longer an anisotropic fluorescent probe the larger the value of $\lambda$ required to orient it.

## 6. CONCLUSION

The studies reported above on the orientation of uniaxially stretched polymer melts show that there is good agreement between the relaxation curve of the average chain orientation determined from IRD and the modulus relaxation curve.

However, the FP results show that, although the $M^2$ dependence of the chain-shrinkage relaxation time predicted by Doi seems to be confirmed qualitatively, it is necessary to improve the slip-link model, first by considering a topological coupling between the chains in the melt, and secondly by taking into account the anisotropy of the strained medium on the orientation of chain segments and its consequence with regard to chain relaxation. At the present time only the topological coupling has been considered.

In the near future, studies by IRD of block polymers with hydrogenated and deuterated sequences should yield a more detailed description of chain relaxation, whereas FP experiments should allow investigation of the effect of polydispersity due to the high sensitivity of the technique which requires a small amount of labelled species.

It clearly appears that FP and IRD techniques, which are easy to apply to oriented polymers either during or after stretching, are powerful tools with which to develop a molecular approach to polymer viscoelasticity.

## REFERENCES

1. DE GENNES, P. G. (1971). *J. Chem. Phys.*, **55**, 572.
2. DOI, M. and EDWARDS, S. F. (1978). *J. Chem. Soc. Faraday Trans. 2*, **74**, 1789, 1802, 1818.
3. DOI, M. (1980). *J. Polym. Sci. Polym. Phys. Ed.*, **18**, 1005.
4. BOWER, D. I. (1981). *J. Polym. Sci. Polym. Phys. Ed.*, **19**, 93.
5. READ, B. E. (1975). In: *Structure and properties of oriented polymers* (Ed. I. M. Ward), Applied Science Publishers, London.
6. JASSE, B. AND KOENIG, J. L. (1979). *J. Macromol. Sci. C*, **17**, 61.
7. MONNERIE, L. (1982). In: *Static and dynamic properties of polymeric solid state* (Ed. R. A. Pethrick and R. W. Richards), D. Reidel, Dordrecht.
8. JARRY, J. P. and MONNERIE, L. (1978). *J. Polym. Sci. Polym. Phys. Ed.*, **16**, 443.
9. FREHLAND, E. (1975). *Z. Naturforsch.*, **30a**, 1241.
10. DESPERS, C. R. and KIMURA, I. (1967). *J. Appl. Phys.*, **38**, 4225.
11. NOBBS, J. H., BOWER, D. I., WARD, I. M. and PATTERSON, D. (1974). *Polymer*, **15**, 287.
12. JARRY, J. P., SERGOT, P., PAMBRUN, C. and MONNERIE, L. (1978). *J. Phys. E, Sci. Instr.*, 702.
13. PINAUD, F., JARRY, J. P. and MONNERIE, L. (1982). *Polymer*, **23**, 1575.
14. NISHIJIMA, Y., ONOGI, Y. AND ASAI, T. (1966). *J. Polym. Sci. C*, **15**, 237.
15. JARRY, J. P. AND MONNERIE, L. (1979). *Macromolecules*, **12**, 927.
16. FAJOLLE, R., TASSIN, J. F., SERGOT, P., PAMBRUN, C. AND MONNERIE, L. (1983). *Polymer*, **24**, 379.
17. LEFEBVRE, D., JASSE, B. and MONNERIE, L. (1983). *Polymer*, **24**, 1241.
18. LODGE, A. S. (1956). *Trans. Faraday Soc.*, **52**, 120.
19. PLAZECK, D. J. (1965). *J. Phys. Chem.*, **69**, 3480.
20. TASSIN, J. F. and MONNERIE, L. (1983). *J. Polym. Sci. Polym. Phys. Ed.*, **21**, 1981.
21. VIOVY, J. L., MONNERIE, L. and TASSIN, J. F. (1983). *J. Polym. Sci. Polym. Phys. Ed.*, **21**, 2427.
22. TASSIN, J. F. (1986). Thèse Docteur-ès-Sciences, Universite Pierre et Marie Curie, Paris.
23. QUESLEL, J. P. (1982). Thèse Docteur-Ingénieur, Universite Pierre et Marie Curie, Paris.
24. JARRY, J. P. (1978). Thèse Docteur-ès-Sciences, Universite Pierre et Marie Curie, Paris.

*Chapter 7*

# ORIENTED POLYAMIDES

A. J. Owen

*Institut für Physik III, Universität Regensburg,
Federal Republic of Germany*

## 1. INTRODUCTION

Polyamides are macromolecular materials containing periodically recurring amide (CONH) groups in the linear molecular chains. They show characteristics which distinguish them from other polymers, due to the amide groups forming intermolecular hydrogen bonds which strongly influence the mechanical behaviour. Polyamides can be divided into two basic types depending on whether the molecules are inherently flexible or rigid:

(1) Aliphatic polyamides or nylons are flexible-chain polymers. The many types of nylon in use differ from each other in the ratio and position of CONH groups relative to $CH_2$ groups in the chains. Nylons are partially crystalline thermoplastics, which can be moulded and extruded to form films or fibres with useful engineering properties.

(2) Rigid-chain aromatic polyamides are finding increasing use in special applications where high temperature stability and high modulus are required and also as liquid crystal polymers in electronic devices. These materials are also synthesised in several chemical species. Aromatic polyamide fibres (aramid fibres) are, however, not thermoplastics, and require strong acids or special solvents for their manufacture. In addition, partially aromatic polyamides (i.e. substances combining aliphatic and aromatic groups) are also of considerable interest.

237

This chapter deals chiefly with relationships between structure and thermomechanical properties of oriented polyamides. Section 2 contains a description of the crystallographic and morphological structure of oriented nylons. In Section 3 the orientation behaviour of nylons is discussed, in particular with respect to load–elongation curves, birefringence and elastic mechanical anisotropy measurements. Thermomechanical properties and special aspects are treated in Section 4. Section 5 then deals relatively briefly with the structure and properties of aromatic polyamides.

## 2.  THE STRUCTURE OF ORIENTED NYLONS

Information on the structure at its various levels is essential for an understanding of the mechanical properties. These aspects will therefore now be considered.

### 2.1. Molecular Structure

The monomeric structure of nylons is summarised in Table 1. A detailed account of the chemistry of nylon synthesis can be found in most general handbooks on polymers. Nylons prepared from an $\omega$-amino acid or by ring-opening polymerisation of a lactam are written with a single suffix ($m$), which gives the number of C atoms per chemical repeat unit, e.g. nylon-6 ($m = 6$), nylon-12 ($m = 12$) etc. Nylons obtained from a diamine and a dibasic acid are written as nylon-$m,m'$. One of the most important of these is nylon-6,6.

Two of the most well known aromatic polyamides, poly($p$-benzamide) (PBA) and poly($p$-phenylene terephthalamide) (PPTA), are also listed for comparison (see also Section 5).

### 2.2. The Crystal Structure of Nylons

Due to the regularity of the molecular chains, nylons form partially crystalline solids. Wide-angle X-ray scattering (WAXS) measurements show that there are two basic equilibrium crystallographic unit cell structures, called the $\alpha$ form and the $\gamma$ form respectively.[1] Table 2 gives the crystal form occurring in ordinary specimens.

The $\alpha$ form is characterised in general by hydrogen bonded planes of fully extended molecular chains, whereas the $\gamma$ form has a shortened chain repeat length due to twisting of the chain necessary to allow complete

## TABLE 1
### MONOMERIC STRUCTURE OF POLYAMIDES

*Aliphatic polyamides* (*nylons*)
Produced by polycondensation of an $\omega$-amino acid or ring-opening polymerisation of a lactam:

Nylon-$m$
$$\left[ \begin{array}{c} \text{N} \\ \text{H} \end{array} \!\!-\!(\text{CH}_2)_{m-1} \!-\! \begin{array}{c} \text{O} \\ \text{C} \end{array} \right]_n$$

Produced by polycondensation of a diamine and a dicarboxylic acid:

Nylon-$m,m'$
$$\left[ \begin{array}{c} \text{N} \\ \text{H} \end{array} \!\!-\!(\text{CH}_2)_m\!-\! \begin{array}{c} \text{H} \\ \text{N} \end{array} \!\!-\! \begin{array}{c} \text{C} \\ \text{O} \end{array} \!\!-\!(\text{CH}_2)_{m'-2}\!-\! \begin{array}{c} \text{O} \\ \text{C} \end{array} \right]_n$$

*Aromatic polyamides* (*aramids*)

PBA
$$\left[ \begin{array}{c} \text{N} \\ \text{H} \end{array} \!\!-\!\!\left\langle \bigcirc \right\rangle\!\!-\! \begin{array}{c} \text{O} \\ \text{C} \end{array} \right]_n$$

PPTA
$$\left[ \begin{array}{c} \text{N} \\ \text{H} \end{array} \!\!-\!\!\left\langle \bigcirc \right\rangle\!\!-\! \begin{array}{c} \text{H} \\ \text{N} \end{array} \!\!-\! \begin{array}{c} \text{C} \\ \text{O} \end{array} \!\!-\!\!\left\langle \bigcirc \right\rangle\!\!-\! \begin{array}{c} \text{O} \\ \text{C} \end{array} \right]_n$$

interchain hydrogen bonding (evidence for dangling hydrogen bonds is lacking[1]).

For the series F nylons, such as nylon-12 and nylon-6 (which will be the main examples quoted in this chapter), the chains have an odd number of CH$_2$ groups between the amide groups; this leads to an alternating sequence of CO and NH groups on a particular side of the chain, and

## TABLE 2
### CRYSTAL STRUCTURE OF NYLONS

| Series | $m$ | $m'$ | Example | Normal crystal form |
|--------|------|------|---------|---------------------|
| A | even | even | nylon-6,6 | $\alpha$ |
| B | even | odd | nylon-6,9 | $\gamma$ |
| C | odd | even | nylon-7,10 | $\gamma$ |
| D | odd | odd | nylon-7,7 | $\gamma$ |
| E | odd | | nylon-11 | $\alpha$ |
| F | even | | nylon-4 | $\alpha$ for $m < 6$ |
| | | | nylon-12 | $\gamma$ for $m > 6$ |

hence to a directionality for the molecule, i.e. neighbouring hydrogen bonded chains are either 'parallel' or 'antiparallel'. The $\alpha$ structure comprises hydrogen bonded planes of antiparallel chains, whereas the $\gamma$ crystal has hydrogen bonding between parallel chains with the hydrogen bonded planes packing in an antiparallel manner.[1,2] This latter structure has a pleated sheet form, with twisting of the chain at the amide groups.

Whereas nylons of series F usually show one or other of the two stable forms, nylon-6 can have either of the two structures.[1,3] Both forms have been observed in a highly annealed film; in specimens which have not been crystallised by annealing, a series of metastable structures can exist which vary continuously in size and perfection from a pseudohexagonal structure to either of the two stable forms.[3] Treatment of nylon-6 with aqueous phenol solution can be used to produce a pure $\alpha$ form, whereas a $\gamma$ structure can be obtained by using iodine and potassium iodide; the mechanical behaviour of oriented films in these two forms will be discussed in Section 3.

In nylon-12, which shows the characteristics of the $\gamma$-form, there is some doubt as to whether antiparallel packing of hydrogen bonded planes of parallel chains occurs in this structure, i.e. evidence in the WAXS pattern for a double basal plane to the unit cell is lacking.[4] This would mean that either all chains within a particular crystal have the same directionality or, as Owen and Kollross[5] suggest, a type of mixed crystal disorder with respect to the chain direction occurs even in well crystallised samples.

In addition, differently oriented samples show different basal planes to the unit cell,[5] suggesting that hydrogen bonded planes parallel to the surface of the sheet occur in uniplanar-axially oriented samples (produced by drawing, rolling and annealing) whereas uniaxially oriented sheets have a spatial hydrogen bonding network, similar to suggestions of Roldan and Kaufman[6] for nylon-6 and Schmidt and Stuart[7] for nylon-8. These points will be discussed further with respect to mechanical properties in Section 3. The reader is referred to standard works (e.g. References 8 and 9) for further details of the crystallographic structure of nylons.

### 2.3. Morphology of Oriented Nylons

Judging from the width of WAXS reflections, nylon crystallites have apparent dimensions of the order of 10 nm. Between the crystallites are transition regions of amorphous or less ordered material. These cause a halo in the WAXS diagram. The arrangement of these crystalline and amorphous regions (morphology) also plays a major part in determining the mechanical behaviour.

The morphology can be investigated by means of transmission electron microscopy (TEM) of stained sections. For example, $OsO_4$ is thought to be reduced by formaldehyde and to react selectively with the carbonamide groups in the amorphous regions of polyamides, with the result that in the electron microscope contrast between crystalline and amorphous regions can be achieved.[10] Other stains, such as $SnCl_2$ and phosphotungstic acid, have also been used.[11,12]

As an example, Plate 1(a) (see p. 251) shows a TEM micrograph of a stained, cold-drawn (draw ratio 3:1) and annealed (170°C) nylon-12 film, which has been sectioned in a plane parallel to the draw direction and perpendicular to the plane of the film.[13] Optical micrographs of the TEM negatives were used to deduce that the draw (and chain) direction is vertical in the diagram. The vertical stripes are microfibrils with a width of about 9 nm. They consist of alternating blocks of crystalline (C) and amorphous (A) material (crystal–amorphous periodicity $\sim 14$ nm). Between the microfibrils are essentially continuous interfibrillar regions of non-crystalline material (I) of approximate width 5 nm; see Plate 1(b), p. 251). The same specimen as in Plate 1(a) showed a two-point meridional small-angle X-ray scattering (SAXS) pattern; see plate 1(c), p. 251. The spacing of the reflections gave a long period of 16·4 nm, which agrees satisfactorily with the crystal–amorphous periodicity along the draw direction seen in the microfibrils.

The absence of equatorial streaking in the SAXS diagram, in conjunction with the TEM image, gives further information about the morphology; in the TEM specimens, the staining material has diffused into the interfibrillar regions and produced contrast between microfibrillar and interfibrillar regions, which would lead to equatorial scattering, as is indeed found in the optical diffraction patterns of the TEM negatives. In the SAXS pattern of uncontrasted material, on the other hand, the absence of equatorial scattering shows that the projection of the electron density on the equator is constant, i.e. the average density of the microfibrils is approximately the same as that of the (uncontrasted) interfibrillar regions.

The half-width of the horizontal intensity profile of the SAXS meridional reflection was found to be approximately independent of position on the meridian,[13] which suggests that the reflection profile is caused chiefly by crystal width effects. A Guinier plot[14] gave a value of 1·9 nm for the radius of gyration ($R_G$) of scattering units from their centre of gravity. If the diffracting elements are of cylindrical cross-section[15] then the effective diameter of the scattering cylinders is $4R_G = 7·6$ nm, which agrees well with the microfibrillar width estimated from electron microscopy.

The observation of a constant width to the meridional reflection for nylon-12 contrasts with the case of polyethylene, where the reflection profile increases in width with increasing scattering angle, and is caused by inclination of the crystal–amorphous interfaces to the draw direction.[16] There are additional, fundamental differences between the morphologies found in polyamides and polyethylene. The above considerations suggest a three-phase model for the structure of drawn nylons, viz. crystal (C), amorphous microfibrillar (A) and amorphous interfibrillar (I), as opposed to a two-phase model. The phase which is continuous in the draw direction is the amorphous I region, unlike e.g. polyethylene and polypropylene which have crystal continuity or crystalline bridges connecting the C regions along the draw direction.[17]

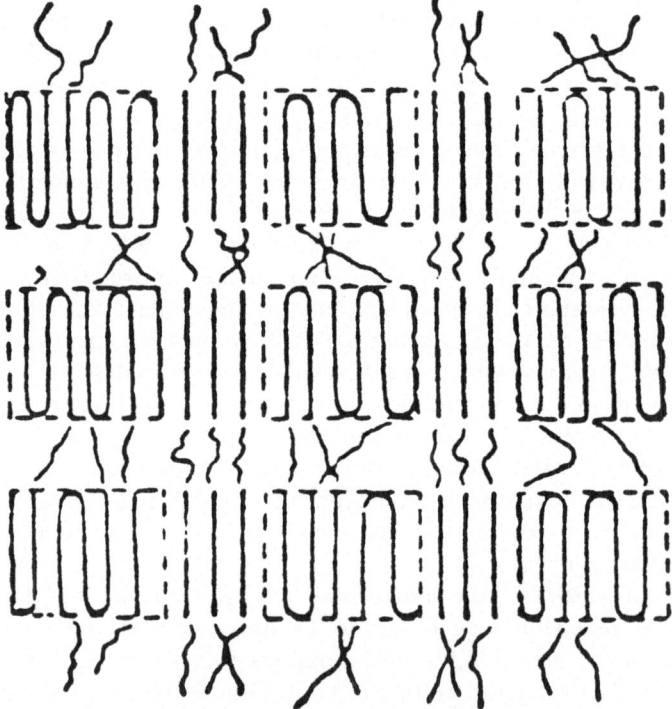

FIG. 1.   Model of microfibrillar fusion via epitaxial crystallisation of extended-chain molecules for nylon-6 fibres (adapted from Reference 19 by permission of the publishers, Chapman and Hall).

Apart from electron microscope evidence for two amorphous phases (A and I) in polyamides, MacKerron showed using IR measurements on deuterated nylon-6,6 that two discrete amorphous components are present, with one of these components having high molecular alignment.[18]

Reimschuessel and Prevorsek[11] observed additional features in their electron micrographs of stained (with $SnCl_2$) nylon-6 fibres. Sections perpendicular to the draw direction indicated the presence of macrofibrils of thickness between about 30 and 100 nm. These were ribbon-like denser areas separated by less dense domains. Both the macrofibrils and the less dense matrix appeared to consist of microfibrils. The macrofibrils were preferentially aligned with their wider surface parallel to the surface of the fibre. They already existed even in the undrawn fibres, becoming thinner in the drawn fibres. According to Prevorsek and coworkers,[19] the crystallites in the microfibrils are arranged in planes perpendicular to the fibre axis. They suggest, as a result of mechanical and diffusive measurements, that crystallites in adjacent microfibrils may be fused together epitaxially (see Fig. 1). The crystalline domains in the microfibrils are presumed to show some chain-folding at the crystal–amorphous boundary, whereas the interfibrillar amorphous material (I) is thought to consist essentially of extended-chain molecules (see Fig. 1).

The structure of aromatic polyamides is to be treated in Section 5.

## 3. THE MECHANISM OF ORIENTATION IN NYLONS

From a scientific point of view, suitably oriented nylon films are of considerable interest, since it may be possible to separate the effects of molecular chain alignment, hydrogen bond alignment and morphology within a single anisotropic film. An understanding of these effects obviously has implications for the industrial use of these materials.

In this chapter we shall concern ourselves mainly with initially isotropic compression-moulded films which are then subsequently oriented by drawing and/or rolling processes. Annealing experiments also play a major part in determining the mechanical properties.

### 3.1. Load–Elongation Curves

The starting point for many investigations of orientation is the measurement of load–elongation curves. Figure 2(a,b) shows typical results for nylon-11 films drawn at different temperatures (adapted from Northolt et al.[20]). Where a peak in the curve occurs, the sample extended by neck formation with subsequent inhomogeneous deformation. Above a certain temperature, homogeneous extension occurs on a macroscopic scale.

On unloading the sample, some of the deformation is reversed, but if the initial elastic region has been exceeded then a 'permanent' deformation will remain, which is both time- and temperature-dependent. This permanent set is accompanied generally by orientation of the structural units of the polymer on all levels.

In Fig. 2(c) a characteristic load–extension curve for a nylon-12 sample is shown (after Reference 21). The region labelled 1 is the elastic range and 2 is the yield point. In range 3 there is inhomogeneous drawing of the initially spherulitic structure; microfibrils are formed in this region, and the crystallites become highly oriented with the chain axes aligned in the draw direction. In region 4 there is a homogeneous drawing of the fibrous structure. Here the fibrils shear relative to each other, the overall amorphous orientation increases gradually and the degree of crystallinity decreases at the expense of the interfibrillar material.[19,22]

The orientation can be monitored by several experimental methods. In the following sections we shall discuss chiefly the birefringence of polarised light and elasticity measurements.

### 3.2. Optical Anisotropy (Birefringence)
For polymers in general there is a correlation between optical birefringence and overall molecular orientation. As an example, the birefringence of cold-drawn sheets of nylon-6 and nylon-12 is shown as a function of draw ratio in Fig. 3 (author's own measurements).

The theoretical curves shown in Fig. 3 are attempted fits according to Ward's aggregate model for a pseudo-affine transformation of 'units' of the structure.[23] The curves labelled 'a' are for a uniaxially oriented sample, whereas curves 'b' refer to a uniplanar-axially symmetric sheet (see Appendix for mathematical relationships). The trend of the results is predicted well by the pseudo-affine model, with both sheets here tending towards being uniplanar-axially symmetric.

### 3.3. Consideration of Two Phases
For annealed samples which show a distinct segregation into crystalline and amorphous regions with a spherulitic morphology in the undeformed state, the orientations of the crystalline and amorphous components may be taken separately into account, since they may be different for the two phases. Leung et al.[24] measured the birefringence ($\Delta n$) of nylon-6 and

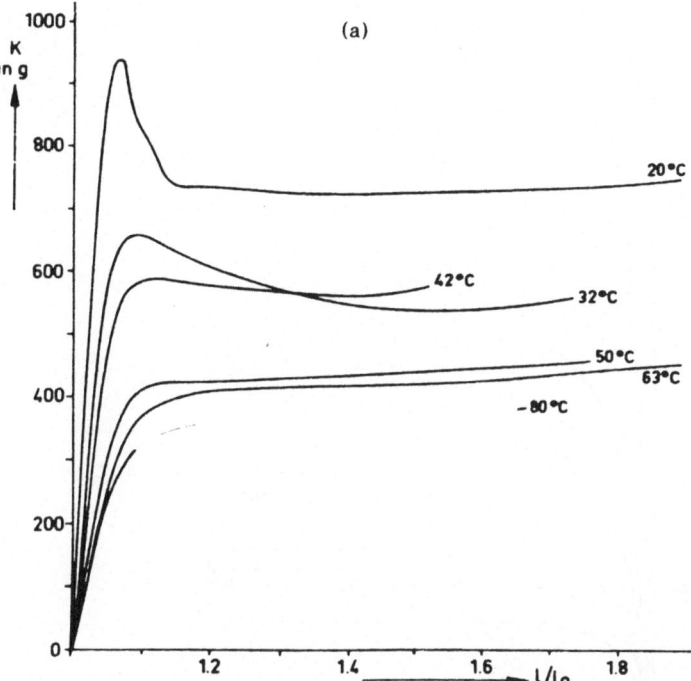

FIG. 2.   Load–elongation curves of nylon-11 drawn at different temperatures: (a) annealed 22°C; (b) annealed 75°C (adapted from Reference 20 by permission of the publishers, Dr. Dietrich Steinkopff Verlag). (c) Load–elongation curve for a nylon-12 sample (after Schnell[21]).

nylon-6,6 as a function of draw ratio (Fig. 4(a)) together with the crystal orientation ($f_c$) from WAXS patterns. Assuming that the birefringence gives a good measure of the total orientation, weighted according to the volume fraction of crystal ($V_c$) and amorphous ($V_a$) phases, then one can write

$$\Delta n = V_c \Delta n_c^\circ f_c + V_a \Delta n_a^\circ f_a \qquad (1)$$

The overall amorphous orientation function ($f_a$) is then found to be as given in Fig. 4(b) for the two polymers. $\Delta n_c^\circ$ and $\Delta n_a^\circ$ are the intrinsic birefringence values for fully oriented crystalline and amorphous regions respectively. (This analysis does not take form birefringence into account.)

### 3.4. Development of Mechanical Anisotropy
The mechanical properties are also related to the orientation of the structure, but in a more complicated fashion than the optical behaviour is.

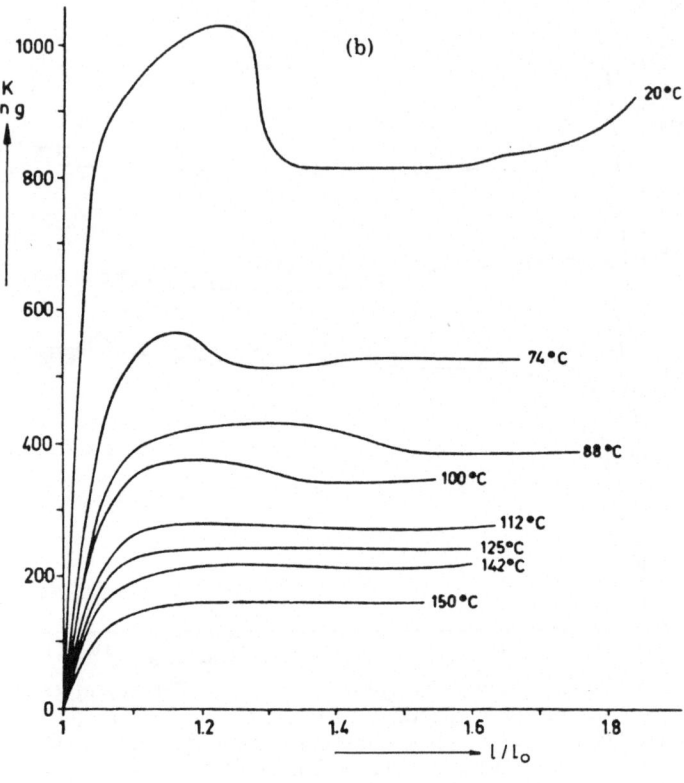

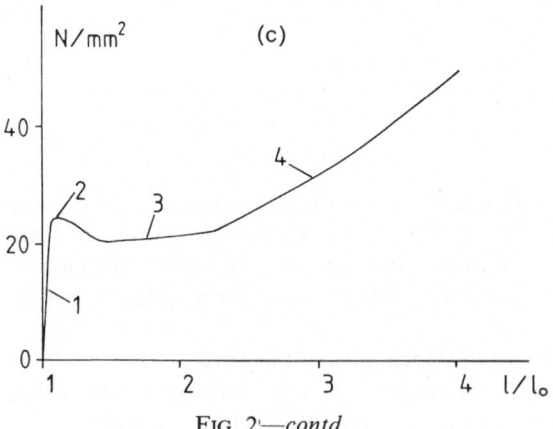

Fig. 2—*contd.*

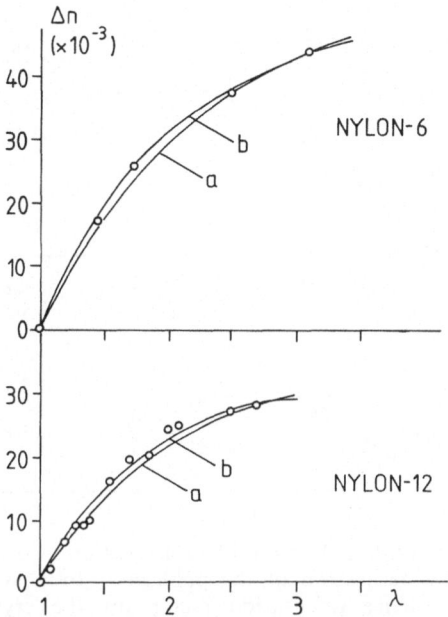

FIG. 3.   Birefringence as a function of draw ratio for cold-drawn nylon-6 and nylon-12. Theoretical fits: (a) pseudo-affine, uniaxial (adapted from Owen and Ward[25] by permission of the publishers, Marcel Dekker); (b) pseudo-affine, uniplanar-axial.

For example, the anisotropic elastic properties relate to averages involving the fourth moment of the orientation distribution. In the plane of a sheet, the tensile compliance $D(\alpha)$ depends on the angle $\alpha$ between loading and drawing direction. Figure 5 shows values of $D(0)$, $D(45)$ and $D(90)$ as a function of draw ratio $\lambda$ for cold-drawn nylon-6 (adapted from Reference 25). The specimen sheets drew homogeneously. Of note is that both $D(0)$ and $D(90)$ first show a small peak at draw ratios of approximately 1·15 and 1·3 respectively, whereafter they fall with increasing draw ratio.

### 3.5. Interpretation of Mechanical Anisotropy in Terms of the Aggregate Model

To analyse the anisotropic mechanical behaviour, it is again useful to adopt Ward's model[23] whereby the polymer is regarded as an aggregate of anisotropic elastic units which are assumed to have elastic properties defined entirely by the components $s_{ij}^{\circ}$ of a compliance tensor. The units

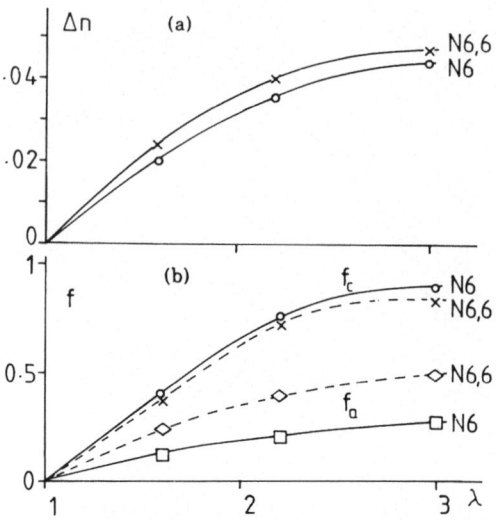

FIG. 4. Results for nylon-6 and nylon-6,6 as a function of draw ratio (adapted from Reference 24 by permission of the publishers, John Wiley & Sons): (a) birefringence; (b) Herman's orientation factor for the crystalline ($f_c$) and amorphous ($f_a$) regions.

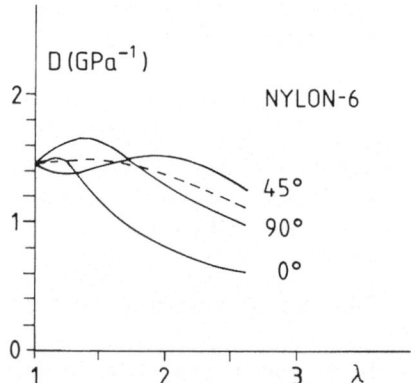

FIG. 5. Compliance anisotropy in nylon-6 as a function of draw ratio. Ten-second compliance at 17·5°C (adapted from Owen and Ward[25] by permission of the publishers, Marcel Dekker).

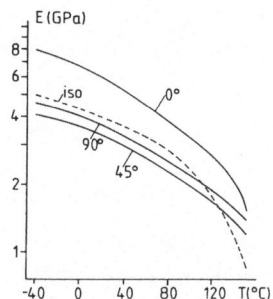

FIG. 6. Tensile modulus anisotropy for nylon-6,6 as a function of temperature (after Reference 24 by permission of the publishers, John Wiley & Sons).

are assumed to retain constant properties but are gradually aligned as the polymer is drawn.

The original aggregate model[23] for essentially a single-phase aggregate has been extended by Samuels and coworkers[26-28] to two phases (crystal and amorphous) under the assumption of uniform stress (Reuss or series averaging).

Using the aggregate model with series averaging, it is quite straight-forward to show that, for any number of phases, $D(0)$, $D(45)$ and $D(90)$ are related to each other by the following invariant:[29]

$$(1/15)D(0) + (8/15)D(45) + (6/15)D(90) = D(\text{isotropic}) \qquad (2)$$

This quantity should be independent of draw ratio according to this model; if, say, $D(0)$ falls with increasing draw ratio then $D(45)$ and $D(90)$ will compensate in such a way that the expression on the left-hand side remains constant.

To illustrate this model, Fig. 5 shows the values of $D(0)$, $D(45)$ and $D(90)$ for cold-drawn nylon-6, together with the invariant (dashed curve). This rises slightly then falls again with increasing draw ratio, showing that in this case the microregions or units change their properties and/or coupling behaviour somewhat with changing draw ratio.

Another example is given in Fig. 6 which shows tensile modulus results for nylon-6,6 plotted as a function of temperature (after Leung et al.[24]). At high temperatures the aggregate of drawn units is generally stiffer than in the isotropic case, whereas at low temperatures the aggregate model describes the behaviour well.

### 3.6. Reduction of the Number of Parameters in the Aggregate Model

Samuels[26] made simplifying approximations in order to reduce the number of unknown parameters in the two-phase aggregate model. With his simplifications, the invariant formulation of this chapter leads to the following expression:

$$D(45) = (1/2)[D(0) + D(90)]$$

or

$$(1/3)D(0) + (2/3)D(90) = D(\text{iso}) \qquad (3)$$

We thus see that this relates to a situation where the $D(45)$ compliance is the average of $D(0)$ and $D(90)$. For nylon samples, we find $D(45)$ always to be relatively large, so Samuels' simplification is not applicable for the case of oriented nylons.

### 3.7. Compliance due to Simple Shear of 'Units'

The relatively large value of $D(45)$ in the fully drawn samples indicates the occurrence of shear on planes parallel or perpendicular to the draw direction.

Neglecting all strain components except simple shear in Ward's single-phase aggregate model leads to the following expressions for $D(0)$, $D(45)$ and $D(90)$ using Reuss averaging:

$$D(0) = (15/2)D_{\text{iso}}\langle \sin^2\theta \cos^2\theta \rangle$$
$$D(45) = (15/8)D_{\text{iso}}(1 - (3/8)\langle \sin^2\theta \rangle - (13/8)\langle \sin^2\theta \cos^2\theta \rangle) \qquad (4)$$
$$D(90) = (15/16)D_{\text{iso}}(3\langle \sin^2\theta \cos^2\theta \rangle + \langle \sin^2\theta \rangle)$$

If the above orientation parameters are calculated as a function of draw ratio using the pseudo-affine scheme, the pattern of anisotropy shown in Fig. 7 is calculated. The actual orientation of cold-drawn nylon-6 at room temperature (cf. Fig. 5) proceeds somewhat differently from that calculated, showing that a simple explanation of all the features of the elastic anisotropy in nylon-6 is not given correctly by this approximation.

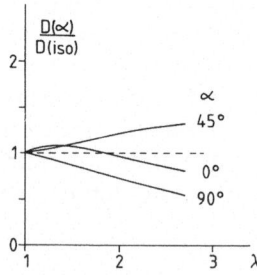

FIG. 7. Theoretically predicted compliance anisotropy: pseudoaffine scheme, Reuss averaging, neglecting all strain components except shear of the units.

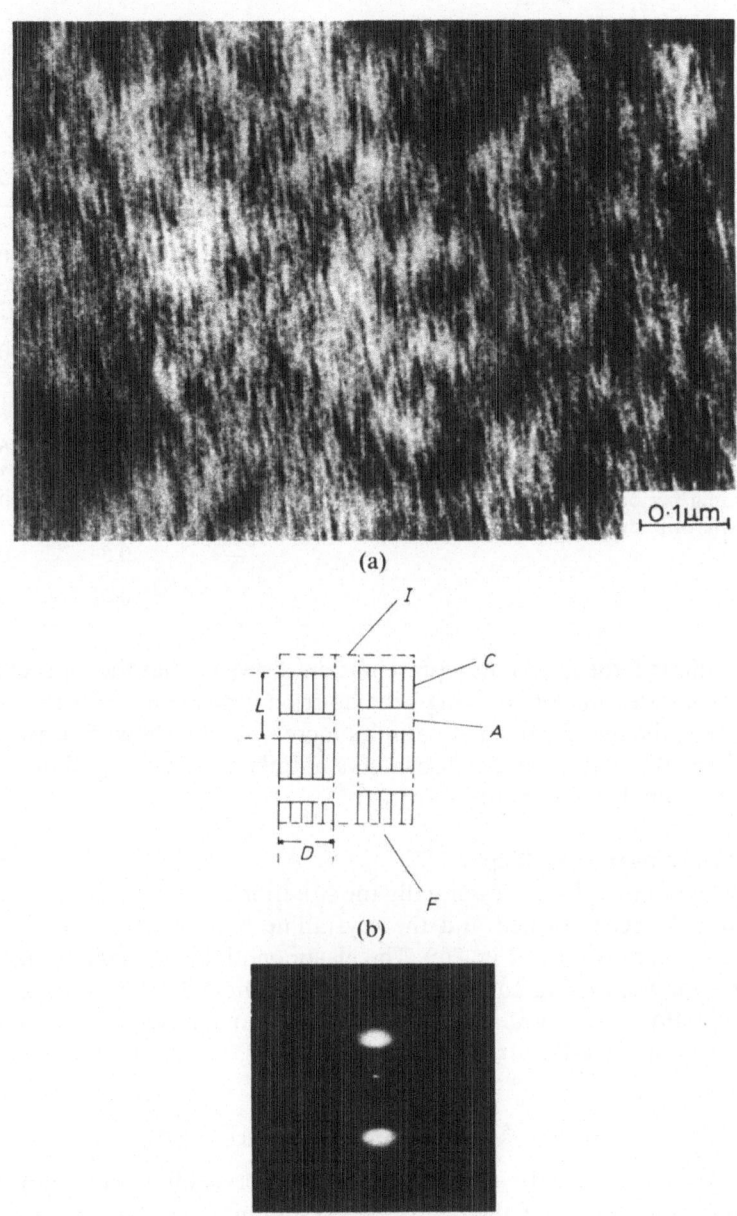

PLATE 1. (a) Electron micrograph of drawn and annealed nylon-12 (draw direction vertical). (b) Schematic model of the structure of oriented nylon-12 (interfibrillar region I; crystal C; amorphous A; microfibril F). (c) SAXS pattern for oriented nylon-12; corresponding to TEM image of (a). (After Geigenfeind and Owen.[13])

TABLE 3
MECHANICAL MODELLING RESULTS

| $\lambda$ | | 1 | 3 | 5·35 |
|---|---|---|---|---|
| (a) | $E$ (GPa) | 2·35 | 4·80 | 5·50 |
| | $\chi$ | 0·63 | 0·56 | 0·34 |
| | $E_c$ (GPa) | 12 | 25 | 25 |
| | $E_a$ (GPa) | 0·7 | 2·34 | 3·26 |
| (b) | $\lambda$ | 0·80 | 0·98 | 0·52 |
| | $\phi$ | 0·787 | 0·570 | 0·654 |
| | $\phi/\lambda$ | 0·98 | 0·58 | 1·26 |
| (c) | Crystal length, $l_c$ (nm) | — | 5·9 | 6·1 |
| | Amorph. length, $l_a$ (nm) | — | 3·0 | 3·2 |
| | Long period, $l_a + l_c$ (nm) | — | 8·9 | 9·3 |
| | Microfib. diam. (or crystal width), $d$ (nm) | — | 11·9 | 7·4 |
| | $l_c/d$ | — | 0·5 | 0·8 |
| (d) | $\zeta/2$ | — | 0·01 | 0·25 |

As mentioned, the invariant representation shows us that the units change their properties and/or coupling behaviour on drawing (at least for nylon-6). In fact, the density of these samples increased slightly with draw ratio, indicating that either an increase in crystallinity or a better packing of the molecular chains occurs on drawing.[25]

### 3.8. Mechanical Modelling

In this section we discuss essentially the situation when a fibrillar structure has already been formed and the crystalline regions are virtually fully oriented, i.e. range 4 in Fig. 2(c). The elastic modulus in such an instance can be analysed using composite mechanical modelling procedures. For example, Prevorsek et al.[19] adapted the Takayanagi model[17] to interpret their modulus results for drawn nylon-6 fibres, where the modulus $E$ is given by

$$E = \lambda[\phi/E_c + (1 - \phi)/E_a]^{-1} + (1 - \lambda)E_a \qquad (5)$$

where $E_c$ and $E_a$ are the respective moduli of crystalline and amorphous domains and $\lambda\phi = \chi$ is the degree of crystallinity. $\lambda = 1$ means that the response of the two phases to an external load is in series (equal stress or Reuss bound) and with $\phi = 1$ there is parallel response (equal strain or Voigt bound). Their results are summarised in Table 3(a, b) for an isotropic and for two drawn fibres. Isotropic material showed a 'mixed' response to

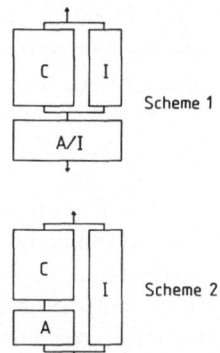

FIG. 8. Coupling schemes used to describe the behaviour of nylon-6 fibres. Scheme 1 is equivalent to Paul's model.[32] Scheme 2, a microfibrillar model. describes the behaviour more satisfactorily. Reproduced by permission of the publishers, Chapman and Hall.

stress with $\phi$ approximately equal to $\lambda$. The fibre of draw ratio 3 showed series coupling ($\lambda = 1$), whereas at higher extension the coupling was again mixed. It is to be noted that $\phi$ and $\lambda$ are not closely linked to the dimensions $l_c$ and $d$ of the crystallites; Table 3(c).

Prevorsek and coworkers[19] additionally used the Halpin–Tsai equation (see e.g. Reference 30):

$$E = E_a(1 + \zeta\eta\chi)/(1 - \eta\chi) \tag{6}$$

with

$$\eta = (E_c/E_a - 1)/(E_c/E_a + \zeta)$$

to analyse their results. They identified the parameter $\zeta$ with the aspect ratio of C regions $2l_c/d$, concluding that C regions behave mechanically like wide platelets, since the factor $\zeta \approx 0$.

Owen[31] analysed the same results using other series-parallel models, taking into account the geometry of a representative volume element (RVE) of the structure. Two schemes were treated (Fig. 8). In Scheme 1 the RVE was divided into infinitesimal slices with normals parallel to the specimen direction. In each slice the strain was assumed to be uniform. The slices were then integrated in series to cover the whole RVE. This is equivalent to Paul's approximate method for the behaviour of certain composite materials.[32] In Scheme 2 the C and A regions of the RVE were coupled in series within a microfibril and the I region was coupled in parallel to the microfibril (see Fig. 8). The analysis showed that Scheme 2 agreed best with the experimental results, i.e. each microfibril should be regarded as a series connection of crystal and amorphous components with relatively little lateral constraint, such that its behaviour approaches the Reuss lower bound. The assembly of microfibrils is interspersed with intermicrofibrillar material, the behaviour of which approaches the Voigt upper bound, as expected for long, parallel constituents.

This interpretation of the results differs from that of Prevorsek and coworkers. The latter attribute the series connection at a draw ratio of 3 to strong interaction between the crystals in neighbouring microfibrils (platelet behaviour); Owen pointed out however that wide platelets in conjunction with an essentially incompressible amorphous region would lead to a high modulus and that the experimentally found low modulus is a manifestation of series coupling with no lateral constraint, which arises if crystals are relatively narrow and do not couple laterally.

The most detailed modelling of oriented nylons has been carried out by Lewis and Ward.[33,34] They studied anisotropic mechanical properties of drawn nylon-6 in both tension and shear for pure $\alpha$ and $\gamma$ phases. In the sheet with $\alpha$ crystals, the chains lay parallel to the draw direction and the hydrogen bonds formed layers parallel to the plane of the sheet. For the sheet with $\gamma$ crystals, the hydrogen bonds now formed layers making an angle of about 60° with the plane of the sheet. Both types of sheet possessed a parallel lamellar morphology.

For the $\alpha$ form, the largest compliance was $s_{55}$ (in our notation), which corresponded to crystallite shear parallel to the chain axis, and could be imagined as the sliding of hydrogen bonded planes over each other in the chain direction. For the $\gamma$ form, where the hydrogen bonded planes are not parallel to the plane of the sheet, this mechanism is not dominant. The reader is referred elsewhere[33-35] for a more detailed account of these investigations.

## 4.   THERMOMECHANICAL PROPERTIES OF ORIENTED NYLONS

In this section some of the time- and temperature-dependent effects which occur in oriented nylons will be described, and the rôle played by the intermolecular hydrogen bonds in determining the thermomechanical properties will be discussed.

### 4.1.   Annealing (Ageing) Effects
In the crystalline regions, the hydrogen bonds are often arranged in a planar configuration, and are supposed to contribute to the strong

cohesion of the crystallites. The bond energy of a covalent bond is about $340 \, kJ \, mol^{-1}$ compared with $20 \, kJ \, mol^{-1}$ for a hydrogen bond in polyamides and about $8 \, kJ \, mol^{-1}$ for a van der Waals bond.[36] In the amorphous regions the hydrogen bonds are responsible for the relatively large water uptake compared to other polymers.

From infrared investigations,[37] virtually all hydrogen bonds are supposed to be formed at room temperature, i.e. there are no ($<1\%$) free amide groups present. On exceeding the glass transition temperature the number of free amide groups increases.[38] If one cools a sample from above to below $T_g$, a portion of these free NH and CO groups will be frozen in. The glass transition which is normally observed at about $50°C$ (in nylon-6) by differential thermal analysis (DTA) is absent immediately after this initial heating process. Only after several days does the DTA transition reappear fully. This is caused by the time-dependent reduction of free NH groups and thereby the time-dependent increase of hydrogen bond concentration. Gordon[39] proposed that the amorphous regions form a network connected by hydrogen bonds, whose order becomes partially destroyed on exceeding $T_g$. The reorganisation of the network on cooling below $T_g$ is determined by the subsequent molecular mobility and the matching up of possible hydrogen bonding sites.

Northolt and coworkers[20] went further and showed that the endothermic transition could be moved to higher temperatures by annealing, and proposed that the weakening of the hydrogen bonded network due to annealing above the transition actually permits an increased cohesion of this network, such that the transition temperature is thereby increased. They showed that there are two major consequences for the deformation behaviour depending on whether the material is at a temperature above or below this transition; drawing a sample below the transition resulted in necking and uniplanar-axial symmetry of oriented films, whereas samples drawn above the transition drew homogeneously and showed uniaxial symmetry; see Fig. 2(a,b).

Kollross and Owen[40] investigated the dynamic mechanical properties at $5 \, Hz$ of oriented nylon-12 sheets which had been subjected to two different thermal treatments. Samples A had been heated to $80°C$ and then stored in a vacuum desiccator at room temperature (for several days); they showed an endothermic transition at about $40°C$ on heating. Samples B had been preheated to $80°C$ (i.e. above the transition) immediately prior to making the mechanical measurements and would show no DTA transition at $40°C$. In this way two sets of samples could be compared with different proportions of associated hydrogen bonds. Figure 9(a,b) shows the

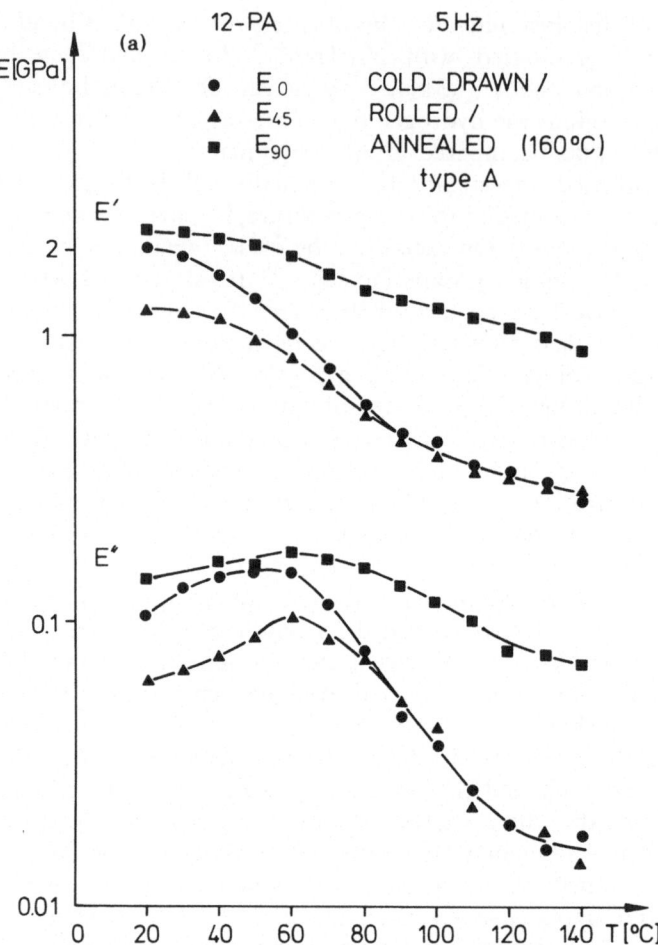

FIG. 9.   Dynamic modulus anisotropy for uniplanar-axially oriented nylon-12: (a) aged; (b) measured directly after heating at 80°C (after Kollross and Owen[40] by permission of the publishers, Butterworth & Co. Ltd).

mechanical anisotropy for sample types A and B for the case of drawn, rolled and annealed nylon-12 (i.e. sheets with uniplanar-axial symmetry). For type A (aged) $E(90)$ is the highest modulus over the whole temperature region, whereas for type B (broken hydrogen bonds) there is a cross-over of $E(90)$ and $E(0)$ below the α transition. Only the $E(90)$ modulus is greatly affected by the thermal treatment; in the aged sample it is about 23% higher than in the immediately measured sample. We interpret the increase

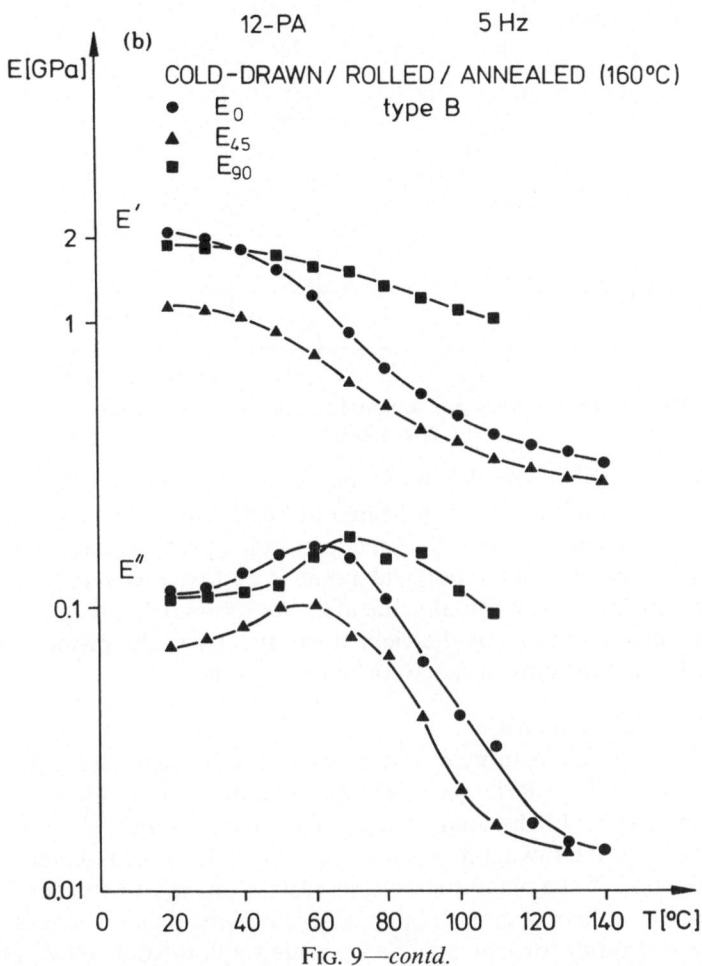

FIG. 9—contd.

in $E(90)$ due to ageing as being the result of the recombination of hydrogen bonds in the amorphous regions. Since the effect manifested itself in the 90° case only, we take this as evidence that the amorphous regions of drawn, rolled and annealed nylon-12 are partially oriented, with the hydrogen bonds aligned preferentially at right angles to the draw direction in the plane of the sheet. The 45° modulus was relatively small in all cases, implying that the shear compliance for shear in the plane of the sheet is generally large.

In Fig. 10 load–elongation curves for a similar nylon-12 are shown

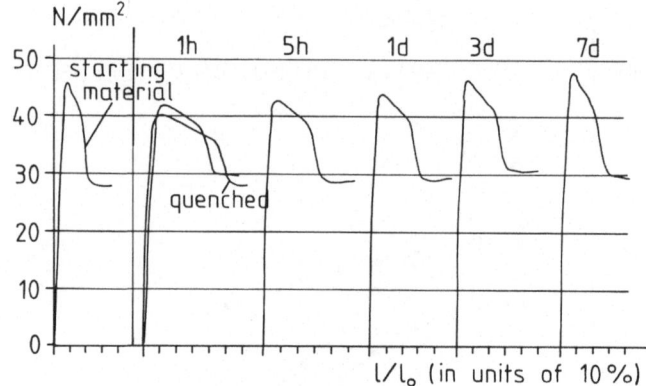

FIG. 10.   Stress–strain curves for nylon-12 after various ageing times (after Wissmann[36]).

(after Wissmann[36]). The left-most curve was obtained for starting material. The samples were then heated at 80°C and quenched in liquid nitrogen. Measurements were then made at indicated time intervals after quenching. It is clear that the yield point is at first lowered, and then gradually returns to its original value after approximately an ageing time of 1 week. Wissmann relates the yield stress directly to the proportion of associated amide groups in the amorphous regions.

## 4.2. Mechanical Relaxations

Mechanical relaxations in nylons in general have been discussed in great detail elsewhere (e.g. Reference 38). The effects of orientation on the dynamic mechanical behaviour of nylon-6,6 are shown in Fig. 11 (results of Leung et al.[24]). Drawing leads to an upshift in the $\alpha$ peak temperature and a reduction in the magnitude of this relaxation, and to smaller drops in the moduli in the relaxation region. Consequently, there is a cross-over in the shear moduli for isotropic and oriented nylon-6,6 at 60°C, above which the value for the oriented sample is much higher. This feature reflects the lowering of segment mobility in the amorphous regions resulting from chain orientation and the presence of taut tie molecules. On the other hand, orientation has little effect on the $\gamma$ process which involves localised motions. Its influence on the $\beta$ process depends on the mode of deformation, with axial tensile measurements giving a large effect.

## 4.3. Improving the Modulus of Nylons

One of the most exciting developments in polymer science in recent years has been the discovery that the elastic modulus of polyethylene fibres can

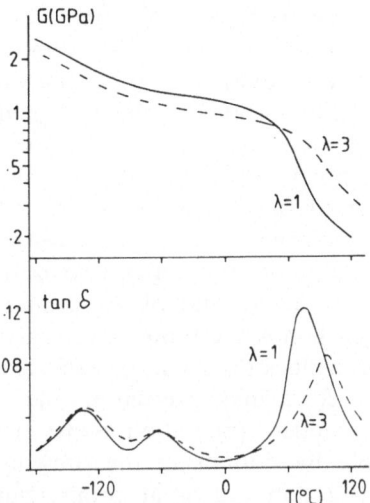

FIG. 11. Temperature dependence of the axial shear modulus $G$ and loss tangent for dry, isotropic and oriented nylon-6,6 (adapted from Reference 24 by permission of the publishers, John Wiley & Sons).

be made extremely large, approaching the theoretical value which lies between 182 and 377 GPa for a stretched $CH_2$ molecular chain, by finding the optimum orientation conditions.[41]

For material oriented in the solid state, Ward and coworkers have shown that a correlation between elastic modulus and draw ratio exists.[42] New extrusion and drawing techniques by Porter and coworkers[43] have resulted in ultra-high modulus polyethylene samples being achieved with a maximum modulus of 222 GPa and a draw ratio of 250. It is of course of considerable fundamental and practical interest to try to obtain equivalent high modulus nylon fibres. Such samples would probably have superior transverse properties to polyethylene, if they could be produced, since the nylon chains are bonded transversely by the relatively strong and stiff hydrogen bonds.

Kaji and Sakurada[44] found from crystal strain measurements a modulus of 164·8 GPa for the chain direction in nylon-6 $\alpha$ crystallites, whereas the $\gamma$ form had a chain modulus of only 20·6 GPa. The reason for this difference lies in the fact that the chain in the $\alpha$ form is properly extended whereas in the $\gamma$ form the molecule is slightly twisted. For this reason it is therefore suggestive to work with polyamides in the $\alpha$ crystalline form.

Attainment of the necessary high draw ratios, however, for polyamides runs into difficulties. Nylon-6 films drawn at room temperature typically can be drawn to a draw ratio of about 3. Fibres may be drawn to perhaps $\lambda = 5$. At room temperature the amide groups in neighbouring chains are associated due to hydrogen bonding. This hydrogen bond network is apparently difficult to break down mechanically and thus prevents high draw ratios being achieved. In polyethylene, on the other hand, the chains shear past each other relatively easily by a $c$-shear mechanism.

Hydrogen bonds in the amorphous regions can be dissociated by raising the drawing temperature above that of the primary (or $\alpha$) relaxation. However, the hydrogen bonds are thought to remain associated in the crystalline regions and to block the occurrence of $c$-axis shear. The crystals then act as physical cross-links preventing the amorphous regions attaining more than a modest orientation. Some increase in modulus is nevertheless achievable by optimising the drawing temperature. For example, wide nylon-12 film drawn at room temperature showed a maximum draw ratio of only 2·5 and a modulus of 1·3 GPa, whereas drawn at 170°C the draw ratio achieved was 4·4 and the modulus (at room temperature) was 3·2 GPa. Samples of nearly square cross-section drew to $\lambda = 4$ at room temperature with a modulus of 3·2 GPa; drawn at 90°C a draw ratio of 5 was obtained with a modulus of 5·3 GPa (Fig. 12 and Reference 45).

Perkins[46] processed oriented nylon thin films by the split billet extrusion technique. The nylon-6 films were plasticised with anhydrous

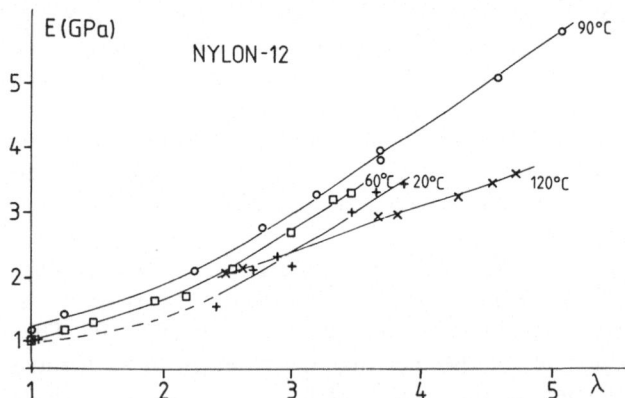

FIG. 12.   Dynamic tensile modulus of nylon-12 at room temperature as a function of draw ratio and drawing temperature.

ammonia prior to extrusion in a high density polyethylene split billet. The coextrusions were carried out with conical dies in the temperature range 65–95°C. In this way films with a tensile modulus of approximately 13 GPa (at room temperature) were achieved.

Acierno and coworkers[47,48] attempted to improve the modulus of nylon-6 by the use of inorganic salts. Some salts alter the properties of polar molecules by a direct binding mechanism. Both anions and cations can be ranked according to their order of effectiveness, i.e. $Li^+$ and $Br^-$ ions are very effective while $K^+$ and $Cl^-$ exhibit essentially no specific interaction. Thus, KCl appears to have no effect on the bulk properties of nylon-6 whereas LiCl and LiBr are quite effective. Addition of small amounts (1–4 wt%) of LiCl to nylon-6 causes a large depression in the melting temperature and crystallisation rate. There is a corresponding increase in the melt viscosity due to the formation of labile network junctions which result from multiple polymer–salt interactions. The presence of these junctions and of small crystalline nuclei resulting from a reduction in crystallisation rate allows the achievement of a relatively high degree of amorphous orientation on deformation of the polymer as, for example, in the extrusion process. Samples can thus be oriented in essentially an amorphous state, and then in principle the salt can be removed by annealing and washing to allow crystallisation in the oriented state. The result using LiCl in nylon-6 was a maximum improvement in modulus for dried material from 3 to 13 GPa.

Richardson and Ward[49] also investigated in detail the behaviour of nylon-6/LiCl mixtures for various spinning, drawing and annealing conditions. A modulus of 8–9 GPa (at room temperature) was obtained for drawn, salted fibres ($\lambda = 6$) compared with $\sim 5$ GPa for drawn, unsalted material. Dynamic mechanical measurements showed that there was reduced chain mobility in the amorphous regions of salted material at room temperature, suggesting strong polymer–ion interactions. They proposed that the modulus increase was caused by an enhanced stiffening effect of extended molecules in these non-crystalline regions. Removal of the salt by washing the fibres in boiling water gave, however, a significant reduction in moduli.

The same approach was taken by Klingerbeck[50] for nylon-12. This polymer was chosen because the amount of hydrogen bonding is considerably less than for nylon-6; however, the disadvantage of nylon-12 is that the ultimate modulus is relatively low ($\sim 21$ GPa). The addition of 3% LiCl brought about a 20–40% increase in the modulus of cold-drawn nylon-12, but washing out the salt in boiling water caused reversal of the

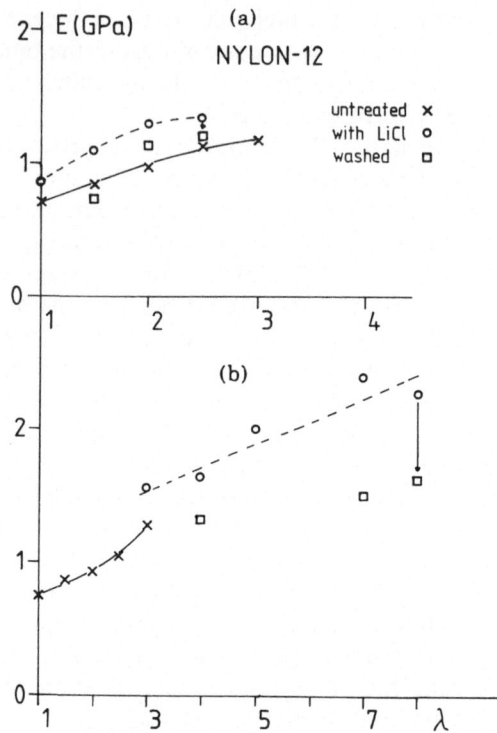

FIG. 13. Modulus of nylon-12 as a function of draw ratio for untreated samples, samples treated with LiCl, and samples which had been subsequently washed and dried: (a) cold-drawn material; (b) hot-drawn (110°C) material.

improvement; see Fig. 13(a). Under conditions of hot drawing (optimum temperature 110°C) a considerable improvement in the maximum draw ratio was achieved ($\lambda = 8$ instead of 2·5), with a corresponding modulus increase (Fig. 13(b)), but again the final modulus after washing out the salt was disappointing. It appears that, as soon as moisture enters the samples again, the improvement in modulus is practically nullified, as found for the case of nylon-6.[47,49]

# 5.   RIGID CHAIN POLYAMIDES

## 5.1. The Structure of Aramid Fibres
It is not intended here to discuss the structure of aromatic polyamide fibres in detail, but only to give some indication of the differences between

nylons and aramids. The present state of knowledge is reviewed in References 43 and 51–53.

A simple picture of an aramid fibre is one of rigid rod molecules held together laterally by hydrogen bonds between the carbonyl and amino moieties of the amide groups. It has been estimated[51] that about 36 monomers would be needed to make a single chain fold, so folding is unlikely. SAXS patterns for PBA and PPTA (see Table 1) show no meridional reflection, only equatorial scattering, so a microfibrillar structure occurs. However, electron microscopy of etched samples indicates that the molecules are organised into extended-chain lamellar structures of high crystallinity. The rod-like molecules are well oriented due to the high persistence length of the chains. The lamellae, with a width of about 600 nm, are arranged in the fibrillar structures which are relatively loosely connected by so-called tie points.[54] Between the lamellae of thickness 35 nm are defect bands or zones which can be seen by electron microscopy in fibres etched with HCl. The WAXS correlation length in the fibre direction is about 80 nm, so it seems that some chains go straight through the defect zones. According to Panar et al.,[54] the defect zones may contain about 50% of the chain ends.

On a much larger scale there appears to be evidence[55] for a pleated-sheet structure in PPTA fibres, with a pleat periodicity of about 500 nm in the fibre direction, where each pleat is deflected by 6° from the fibre direction. These pleats can be observed by polarisation light microscopy, and by scanning and transmission electron microscopy under suitable conditions.

In addition, unlike the situation in conventional nylon fibres, the hydrogen bonding direction appears to lie radially in the aramid fibres.

## 5.2. Some Properties of Aramid Fibres

Aromatic polyamides cannot be melt spun; they decompose at high temperatures. Thermal analysis shows decomposition at about 580°C with no other features.

Hydrogen bonding solvents or strong acids are used to prepare spinning solutions or dopes. They form optically anisotropic solutions in both amide and acid solvent systems, i.e. lyotropic liquid crystals. The liquid crystal type involved is thought to be nematic, which is the simplest form of liquid crystal and involves parallel orientation of large groups of the rod-like molecules. The molecules in the liquid crystal phase have short range order and can be readily oriented in the extensional flow produced by passing through a spinneret. The relaxation rate to a disordered state is slower than the coagulation of the polymer, and so a fibre of highly oriented rigid chains can be produced. Once dried, the fibres are generally not redrawable. They can be heat-treated for short times to 700°C and for long times to 400–500°C under tension to increase the modulus and breaking strength. It is thought that heat treatment permits the chains to slip over one another until hydrogen bonds are brought into register and regions of crystalline order are formed.

The mechanical properties already achieved for aramid fibres have opened up a new era in fibre technology. They possess the strength, modulus and thermal stability of even some glass or steel fibres, yet they have the advantage of low density with good fatigue and abrasion resistance. Applications include many uses as a reinforcing fibre in composites. Kevlar (DuPont) aramid fibres can be used as cut fibre, continuous filament or as pulp, e.g. to replace asbestos in brake linings. The theoretical modulus of PBA and PPTA is approximately 200 GPa; values in the region of 60 and 120 GPa are typical for commercial Kevlar fibres.[51]

### 5.3. Composites based on Aromatic and Aliphatic Polyamides

Of great potential use is also the possibility of combining the advantage of aliphatic and aromatic components. Takayanagi[56] has reported on the properties of several blends of rigid-rod polyaramides or their block copolymers and nylon-6 or nylon-6,6 as matrix polymers. For a blend of 7% PPTA ($M_v = 34\,000$) in nylon-6, 30 nm diameter evenly dispersed 'needles' of PPTA are found, whereas for $M_v = 4500$ the needle diameter is 15 nm. Extrusion of the blend results in orientation of the needles; the nylon matrix, on the other hand, remains unoriented. Under certain growth conditions, aliphatic folded-chain nylon crystals could be grown on the needles of PPTA.

A small addition of PPTA to nylon-6 causes a marked increase in modulus and strength. Extensibility is improved by chemical bonding, i.e. for a copolymer as opposed to a blend. Morphologically this leads to finer microfibrils and a stronger interface between needles and matrix. It appears that the reinforcing effect achieved by addition of PPTA or PBA is very high compared with conventional (e.g. glass) fibre reinforced materials.

### 5.4. Concluding Remarks

Only the future will show us how polyamides will fare with respect to other substances in the very competitive world of materials technology. It is clear that aliphatic and aromatic polyamides belong to different generations of polymeric material, and they consequently have different properties and uses. Nevertheless, both types have been treated in this chapter, since they both have hydrogen bonding as a common denominator. The choice of subject matter in this chapter has been a personal one, mainly deriving from a basic scientific interest in the orientation process and the effects which hydrogen bonding has on the behaviour.

## APPENDIX

Drawn fibres with cylindrical cross-section are uniaxially symmetric (transversely isotropic); $\lambda_1 = \lambda_2$ in Fig. A1. For drawn film or sheet material it is often observed that, when the sample width greatly exceeds the sample thickness, and the grip separation in the testing equipment is relatively small, the relative strain at right angles to the draw direction is

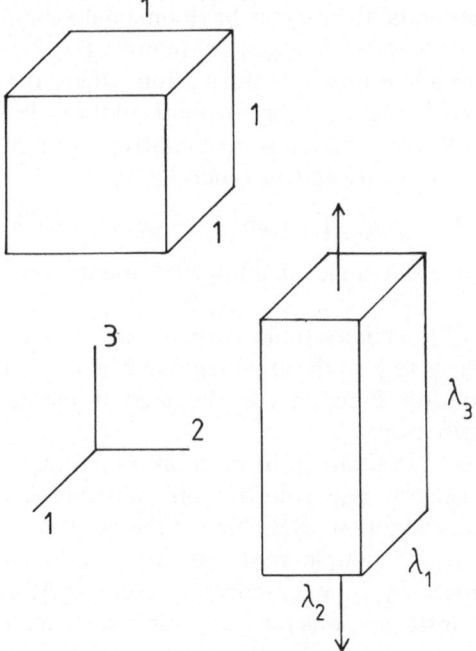

FIG. A1.   Extension ratios $\lambda_1$, $\lambda_2$, $\lambda_3$ for a general extension of a unit cube of material; '3' is the extension direction.

TABLE A1

| $\lambda$ | $\Delta n/\Delta n_{max}$ | |
|:---:|:---:|:---:|
| | Uniaxial $(1/\sqrt{\lambda}, 1/\sqrt{\lambda}, \lambda)$ | Uniplanar-axial $(1, 1/\lambda, \lambda)$ |
| 1·0 | 0·000 | 0·000 |
| 1·5 | 0·255 | 0·287 |
| 2·0 | 0·431 | 0·461 |
| 2·5 | 0·551 | 0·568 |
| 3·0 | 0·637 | 0·638 |
| 3·5 | 0·699 | 0·686 |
| 4·0 | 0·746 | 0·722 |
| 4·5 | 0·782 | 0·749 |
| 5·0 | 0·811 | 0·769 |

not the same in the directions parallel and perpendicular to the plane of the film (i.e. $\lambda_1 \neq \lambda_2$). This also applies to rolled samples.

The aggregate model can be extended to cover a general deformation. Assume that the 'units' themselves have uniaxial symmetry, but that the sample as a whole deforms in a general manner ($\lambda_1$, $\lambda_2$, $\lambda_3$; Fig. A1). Also assume a pseudo-affine type transformation, where the rod-like units of the material rotate in the same way as points in the bulk sample. Using the nomenclature of Reference 23, it is then relatively straightforward to show that the birefringence $\Delta n$ is given generally by

$$\Delta n = \Delta n_{max}(1 - \langle \sin^2 \theta \rangle - \langle \cos^2 \phi \sin^2 \theta \rangle) \tag{A1}$$

where $\theta$ and $\phi$ are the angles defining the orientation of the unit in the aggregate.

For the case of uniaxial sample symmetry, $\phi$ is constant for each unit and $\cos^2 \phi$ averages to $\frac{1}{2}$, so the expression in brackets reduces to the usual Hermans orientation function for the second moment of a uniaxial orientation distribution.

When it comes to evaluating the orientation parameters in eqn (A1) for the completely general case, considerable mathematical difficulties arise. Consequently, we restrict ourselves here to the ideal case of uniplanar-axial orientation where the sample width remains unaltered ($\lambda_2 = 1$) and the thickness is reduced by $1/\lambda = \lambda_1$ (constant volume). It can then be shown that the angle transformations for such a uniplanar-axial deformation are given by

$$\tan \theta = \frac{1}{\lambda^2} \sqrt{\left(\frac{1 + \lambda^2 \tan^2 \phi'}{1 + \tan^2 \phi'}\right)} \tan \theta'; \quad \tan \phi = \lambda \tan \phi' \tag{A2}$$

$\phi$ now changes with draw ratio, and the change in $\theta$ depends on the azimuthal angle $\phi$.

An analytical calculation of $\Delta n$ as a function of $\lambda$ from eqns (A1) and (A2) involves difficult integrals. Consequently, we made a numerical calculation by computer of the orientation parameters as a function of draw ratio, in order to evaluate $\Delta n(\lambda)$. These calculated values of $\Delta n$ are shown in Table A1 compared with the uniaxial case. They are also plotted as curves 'b' in Fig. 3. It can be seen that uniplanar-axial orientation gives a slightly more rapid increase in birefringence (in the plane of the sheet) at small extension ratios than is given by uniaxial orientation. Above a draw ratio of 3 the uniaxial case has the higher birefringence (for the same maximum value in each case). The fit with the results for both nylon-6 and nylon-12 is now very good, in agreement with the observation that the sample sheet did show some biaxial component to the deformation (not measured quantitatively).

## REFERENCES

1. KINOSHITA, Y. (1959). *Makromol. Chem.*, **33**, 1.
2. ARIMOTO, H. (1964). *J. Polym. Sci.*, *A2*, **3**, 2283.
3. PARKER, J. P. and LINDENMEYER, P. H. (1977). *J. Appl. Polym. Sci.*, **21**, 821.
4. NORTHOLT, M. G., private communication.
5. OWEN, A. J. and KOLLROSS, P. (1983). *Polym. Commun.*, **24**, 303.
6. ROLDAN, L. G. and KAUFMAN, H. S. (1963). *Polym. Lett.*, **1**, 603.
7. SCHMIDT, G. F. and STUART, H. A. (1958). *Z. Naturforsch.*, **13a**(3), 222.
8. BONART, R. and ORTH, H. (1976). In: *Ullmanns Encyklopädie der technischen Chemie*, Band 11 Fasern, Struktur, p. 205.
9. *Kunststoff Handbuch* (1966). Band 6–Polyamide, Karl Hanser Verlag.
10. WEBER, G., KUNTZE, D. and STIX, W. (1982). *Colloid Polym. Sci.*, **260**, 956.
11. REIMSCHUESSEL, A. C. and PREVORSEK, D. C. (1976). *J. Polym. Sci. Phys.*, **14**, 485.
12. HESS, K., GÜTTER, E. and MAHL, H. (1960). *Kolloid-Z.*, **168**, 37.
13. GEIGENFEIND, R. and OWEN, A. J. (1985). *Colloid Polym. Sci.*, **263**, 116.
14. GUINIER, A. (1963). *X-ray diffraction*, W. H. Freeman & Co.
15. BOLDUAN O. E. A. and BEAR, R. S. (1951). *J. Polym. Sci.*, **6**(3), 271.
16. BLÖCHL, G. and OWEN, A. J. (1984). *Colloid Polym. Sci.*, **262**, 793.
17. TAKAYANAGI, M., IMADA, K. and KAJIYAMA, T. (1966). *J. Polym. Sci.*, *C*, **15**, 263.
18. MACKERRON, D. H. (1985). *Polym. Commun.*, **26**, 131.
19. PREVORSEK, D. C., KWON, Y. D. and SHARMA, R. K. (1977). *J. Mater. Sci.*, **12**, 2310.
20. NORTHOLT, M. G., TABOR, B. J. and VAN AARTSEN, J. J. (1976). *Progr. Colloid Polym. Sci.*, **57**, 225.
21. SCHNELL, H., personal communication.
22. PETERLIN, A. (1975). *Colloid Polym. Sci.*, **253**, 53.

268                                      A. J. OWEN

23. WARD, I. M. (1962). *Proc. Phys. Soc.*, **80**, 1176.
24. LEUNG, W. P., HO, K. H. and CHOY, C. L. (1984). *J. Polym. Sci.*, **22**, 1173.
25. OWEN, A. J. and WARD, I. M. (1973). *J. Macromol. Sci.*, **B7**(2), 279.
26. SEFERIS, J. C. and SAMUELS, R. J. (1979). *Polym. Eng. Sci.*, **19**, 975.
27. SEFERIS, J. C., MCCULLOUGH, R. L. and SAMUELS, R. J. (1977). *J. Macromol. Sci.-Phys.*, **B13**, 357.
28. SEFERIS, J. C., MCCULLOUGH, R. L. and SAMUELS, R. J. (1976). *Polym. Eng. Sci.*, **16**, 334.
29. OWEN, A. J., *Colloid Polym. Sci.*, submitted for publication.
30. JONES, R. M. (1975). *Mechanics of composite materials*, McGraw-Hill, p. 114.
31. OWEN, A. J. (1981). *J. Mater. Sci.*, **16**, 2324.
32. PAUL, B. (1960). *Trans. Metall. Soc. AIME*, **218**, 36.
33. LEWIS, E. L. V. and WARD, I. M. (1980). *J. Macromol. Sci.-Phys.*, **B18**, 1.
34. LEWIS, E. L. V. and WARD, I. M. (1981). *J. Macromol. Sci.-Phys.*, **B19**, 75.
35. WARD, I. M. (1983). *Mechanical properties of solid polymers*, 2nd Edn, John Wiley.
36. WISSMANN, R. (1980). Doktorarbeit, TU Berlin.
37. TRIFAN, D. S. and TERENZI, J. F. (1958). *J. Polym. Sci.*, **28**, 443.
38. MCCRUM, N. G., READ, B. E. and WILLIAMS, G. (1967). *Anelastic and dielectric effects in polymeric solids*, Wiley.
39. GORDON, G. A. (1971). *J. Polym. Sci. A2*, **9**, 1693.
40. KOLLROSS, P. and OWEN, A. J. (1982). *Polymer*, **23**, 829.
41. BARHAM, P. and KELLER, A. (1975). *J. Polym. Sci. Lett.*, **13**, 197.
42. CAPACCIO, G. and WARD, I. M. (1974). *Polymer*, **15**, 223.
43. PORTER, R. S. (1984). In: *High modulus polymers and composites* (Ed. C. L. Choy), Chinese University Press, Hong Kong.
44. KAJI, K. and SAKURADA, I. (1974). *J. Polym. Sci.-Phys.*, **12**, 1491.
45. BAUER, K.-H. (1984). Diplomarbeit, University of Regensburg.
46. PERKINS, W. G. (1978). PhD Thesis, University of Massachusetts.
47. ACIERNO, D., LA MANTIA, F. P., POLIZZOTTI, G., AFFONSO, G. C. and CIFERRI, A. (1977). *Polym. Lett.*, **15**, 323.
48. ACIERNO, D., LA MANTIA, F. P., TITOMANLIO, G. and CIFERRI, A. (1980). *J. Polym. Sci.-Phys.*, **18**, 739.
49. RICHARDSON, A. and WARD, I. M. (1981). *J. Polym. Sci. Polym. Phys. Ed.*, **19**, 1549.
50. KLINGERBECK, M. (1982). Thesis, University of Regensburg.
51. CARTER, G. B. and SCHENK, V. T. J. (1975). In: *Structure and properties of oriented polymers* (Ed. I. M. Ward), John Wiley, Applied Science Publishers.
52. WARD, I. M. (1982). *Developments in oriented polymers—1*, Applied Science Publishers, p. 158.
53. CIFERRI, A. and WARD, I. M. (1979). *Ultra high modulus polymers*, Applied Science Publishers.
54. PANAR, S., AVAKIAN, P., BLUME, R. C., GARDNER, K. H., GIERKE, T. D. and YANG, H. H. (1983). *J. Polym. Sci. Polym. Phys. Ed.*, **21**, 1955.
55. DOBB, M. G., JOHNSON, D. J. and SAVILLE, B. P. (1977). *J. Polym. Sci. Polym. Phys. Ed.*, **15**, 903.
56. TAKAYANAGI, M. (1980). *J. Macromol. Sci.-Phys.*, **B17**, 591.

# INDEX